Sound in Indian Film and Audiovisual Media

Sound in Indian Film and Audiovisual Media

History, Practices, and Production

Budhaditya Chattopadhyay

Routledge
Taylor & Francis Group
LONDON AND NEW YORK

The publication of this book is made possible by a grant from Swiss National Science Foundation.

First published in 2024 by Amsterdam University Press Ltd.

Published 2025 by Routledge
4 Park Square, Milton Park, Abingdon, Oxon OX14 4RN
605 Third Avenue, New York, NY 10158

Routledge is an imprint of the Taylor & Francis Group, an informa business

ISBN: 9789463724739 (hbk)
ISBN: 9781041186397 (pbk)
ISBN: 9781003704058 (ebk)
NUR 670

Cover illustration: Filmstill from Teen Kanya (Satyajit Ray, 1961), image courtesy NFDC - National Film Archive of India.

Cover design: Coördesign, Leiden

DOI: 10.5117/9789463724739

For Product Safety Concerns and Information please contact our EU representative:
GPSR@taylorandfrancis.com
Taylor & Francis Verlag GmbH, Kaufingerstraße 24, 80331 München, Germany

Table of Contents

Acknowledgements

This book was developed from my research conducted at the Faculty of Humanities, University of Copenhagen, Denmark, from 2012 to 2015, and at the Academy of Creative and Performing Arts (ACPA), Leiden University, the Netherlands, from 2016 to 2017. I thank the Faculty of Humanities at the University of Copenhagen, Denmark, for supporting and financing the research and fieldwork. Later, further research and fieldwork for this book was supported by the Satyajit Ray Film and Television Institute, Ministry of Information and Broadcasting in India during a Film Research Fellowship (2017–18). I would like to thank SRFTI for the support. Thanks also to the National Film Archive of India, Pune for hosting me during a research visit in 2019. A Mellon Fellowship at the Centre for Arts and Humanities, American University of Beirut from 2018 to 2019 also helped me further develop the manuscript, providing me with productive time for continuing to write. The manuscript was completed between the spring of 2022 and spring of 2023, thanks to the support of the Critical Media Lab, Institute Experimental Design and Media Cultures, Basel, University of Applied Sciences and Arts Northwestern Switzerland (FHNW), where I am currently working as a professor with a practice-to-science research grant received from the Swiss National Science Foundation (SNSF).

I intervene in the ongoing debates in film sound research, and in the history, aesthetics, and production of what is known as 'Indian Cinema', with the perspective of a sound practitioner. This practice-led intervention has allowed me to make critical observations and self-reflection in order to delineate the historical accounts of developments accommodating in-depth analysis of sound from the personal experience of the practitioner. Firstly, I would like to acknowledge the insights drawn from fellow sound practitioners and the filmmakers trying to push the margins of sound practice in India throughout its history. Their vital contributions have enriched the development of experience in film and audiovisual production. The book would not have been possible to realize without the knowledgebase of many eminent sound practitioners in India. Thanks to Sarah Waring for her very helpful editorial suggestions and the collaboration. Lastly, I would like to thank my mother Shipra Chatterjee and the family, and some of my true-blue friends across the globe for always being there for me in times of difficulty.

1. Introduction

Abstract: Chapter 1 introduces the historical, conceptual, methodological, and theoretical framework of research into Indian film and media arts sound production. Setting the book's scholarly tone, several concepts are discussed as key research entry points. Critically listening to the trajectories of sound production in India, a hypothesis is proposed exploring technological innovations and shifts that have led to manifold sound recording and presentation advances through various production phases. In this chapter, historical trajectories are studied to understand the shifts in various differing yet concurrent sound practices used to engage audiences. Critical observation, reflection, and analysis of passages of sound from representative Indian film and audiovisual media works, specifically from the three phases of sound production identified, are made to qualify these claims.

Keywords: Indian cinema, sound studies, realism, media aesthesis, media art history, film sound

Background Sounds

In the autumn of 2006, I was a final-year student of audiography[1] at the Satyajit Ray Film and Television Institute (SRFTI),[2] a prestigious institute for film education in India. I was invited to speak about my experience as an emerging sound art practitioner and my concerns regarding the use of sound in cinema at a conference titled *Sound Cultures in Indian Cinema*.[3] The conference was held with the intention of launching an Indian film

1 There are no academic institutions in India for sound art studies per se. In India, audiography is a field concerned with the study of sound recording and design, primarily in the context of film and music production, but when choosing their personal projects, enthusiastic students can push for artistic freedom and are encouraged to experiment.

2 See: www.cilect.org/profiles/942#.YI_LJSN95-U.

3 See: http://osdir.com/ml/culture.india.sarai.reader/2006-11/msg00001.html.

Chattopadhyay, B., *Sound in Indian Film and Audiovisual Media: History, Practices, and Production*. Taylor & Francis Group, 2024
DOI 10.5117/9789463724739_CH01

sound research programme in the Department of Film Studies at Jadavpur University in Kolkata. I presented a student paper at the conference, which was later published in the *Journal of the Moving Image* (2007).[4] The paper was my first critical intervention within the field of (film) sound research and, as such, allowed me to enter academic debates around sound in film and audiovisual media as an active sound practitioner and artist.

This intervention was significant for several reasons. Firstly, as a sound practitioner, that is, one who explores the medium of sound for aesthetic, cultural, and artistic purposes, as well as one who theorizes about it, I was able to offer novel insights into the ongoing academic debates in film sound research, which, up to that point, had been based on mostly non-concrete ideas about sound recording and design in films by seasoned academics, who were not sound practitioners themselves. Secondly, the process of devising the paper also helped to advance my own development as an artistic researcher and thinker, allowing me to question the established standards of sound practices that I had come to know through my film school education. The technologically deterministic approach[5] that I had to deal with in my own studies as a student of sound production was the area I chose to question and discuss in my student paper. I also confronted the lack of use of site-specific sounds or ambience in Indian films and intended to shed light on how ambience could contribute to the sense of place if given enough scope in filmmaking. I argued that imposed limitations practiced in standard film sound recording and design create lapses in the inclusion and recognition of site-specific sound information and the sonic reality in cinematic space. This critical stance was in response to the screen-centric and visually dominated field of film production. The paper was an initiation into my ongoing efforts to articulate my artistic practice and research creation with sound within a growing interest in Sound Studies, and academia in general. The paper not only aimed to voice my concerns on these issues but also intended to situate my inquiry and practice within this conundrum and in the genealogy of sound practice in the Indian subcontinent prevalent at that time. These questions and polemics were crucial in the conception and development of this book.

4 See: www.jaduniv.edu.in/templates/newpages/ju_journals.html#mi.

5 Film school education in India mainly revolves around cinema technology in order to train students in the skillful operation of the tools and machines necessary for recording and production. This overt emphasis on technology often ignores the aesthetic sensibilities of working with and thinking through, in my case, the subtleties of sound and listening. Students who are sensitive to the artistic potential of sound may take distance from the educational system and create their own practices independent of the intervention of film school mechanisms.

By describing my background, the case that I would like to articulate here is the potential contribution that I can make to existing research in film sound as someone who is actively involved in sound practice, and how this contribution can be seen as novel and relevant to generating new knowledge. My inquiry also drives the search for a comprehensive positioning of the sound practitioner within the discourse of sound in film and audiovisual media.

If I look back on my aforementioned first academic paper, much has changed since its publication in 2007, and these changes are indeed instrumental in posing the pertinent questions I asked at that time, once again, within a new context – the context of a thorough digitalization of contemporary sound production and its ramifications for sonic experiences. Regarding the issue of an alleged underestimation of the field of sound in Indian films (Rajadhyaksha, 2007; Gopalan, 2002; Ganti, 2012), I lamented in the above-mentioned paper that Indian cinema has generally seemed to be hostile to the rich and realistic site-specific sound spectrum on a:

> [M]arried soundtrack[6] even when a film is shot on location. On location, sound is usually controlled to enter film space, and, on a film set, sound is limited to mere voice and sync effects, making the construction of a soundtrack mostly dependent on asynchronous means of sound sourcing such as available stock sound. In the process, film sound, instead of capturing the vibrant site and its spatial dynamics, drifts away from documenting the sound of an original space. Sound-making goes closer to a synthetic design by a sound operator working under the spectre of mechanical craftsmanship; the sound practitioner's religion of spontaneous hearing loses validity. (Chattopadhyay, 2007, p.x)

But the causes for this lamentation did not remain for long. A revolution was already in progress in Indian cinema following the decision, after much deliberation, to shoot the film *Lagaan* ('Land Tax') (Gowariker, 2001) with 'sync' sound made with digital technology.[7] This revolution brought an awareness of how the direct recording of location sound and the live performance of actors are perceived by filmgoers as far more convincing

6 A 'married soundtrack' is the final mix of the soundtrack synchronized with the moving image on the final mix of all the audiovisual information on the cinematic media, known as 'married print' in film sound practitioner's terminology.

7 An abbreviation of synchronized sound recording made on film location, or the film set, gathering mostly the live performance of actors' voice, effects, and ambience, instead of re-creating them inside the studio. This term will be discussed later in detail.

and realistically believable than the techniques of dubbed films of the 1960s, 70s, and 80s that were exemplifying the established industry standards (with the exception of a few independent filmmakers following a different practice). This dawning awareness was significant in fundamentally shifting the spectator's experience of sound, namely in the proliferation of (an apparent) realism, exploring the 'capacity not only to mirror but also to govern the human perceptual apparatus' (Rogers, 2013, p. 32) towards an 'embodied experience'.[8] This sense of realism and embodiment has been achieved, I assume, through a recognition of the specifics of the spatial sensations in the narrative or, more generally, through a sensitivity to the 'site of sound' (LaBelle and Martinho, 2011). If we look at Indian cinema in the period between 2001 and 2009, there were a number of films that embraced this methodology of sync sound or location-specific digital multi-track recording in live synchronization before it gradually became the standard from 2009 onwards. The novel experience of sync sound textures replete with evidence of the site led to the popular appreciation of a revived sense of sonic realism and spatiality. However, this development didn't make my earlier lamentation completely invalid. My previous concerns about a lack of spatial information in the sound of Indian cinema led to new questions: what do these new developments signify, and how can they be understood from a historical perspective? This set of renewed inquiries has inspired me to take up the course of the present project, and it is these broader questions that drive this book.

If we look at the trajectories of sound production within the cinema of pan-India, we will observe that sound has been inconsistently rendered and produced through various phases of production practices under the effects of technological transformations, innovations, and resultant aesthetic shifts. There have been phases of sound practice, such as the entire 'dubbing era' (1960s–1990s) that cared less about reality while giving more importance to typical entertainment tropes such as 'song and dance' sequences. However, there are also phases such as the 'digital era' (2001–) where the 'realistic' and concrete representation of site is observed. The dominant factors that have determined these shifts are rooted in technological developments (Kassabian, 2013; Kerins, 2011, 2006; Lastra, 2000; Altman, 1992), whereas

8 An embodied experience of sound in film and other mediated environments is provided by site-specific sound recordings – including room tones, low frequency rumbles, and other bodily-perceptible ambient sounds – dispersed in a spatial organization following a multi-channel surround sound design. As argued by Mark Kerins (2006, 2011), these practices find prominence in the digital era of sound production. I am using the term in the context of discussing the digital era in Indian cinema to underscore the unique capacity of surround design and sync sound.

certain aesthetic choices were made available by these technological phases shaping the historical evolution of sound practice in Indian cinema. Here, this history is studied and written about in order to understand the various differing and concurrent practices, methodologies, and approaches, as well as the shifts in recording and spatially organizing sound for the purpose of storytelling. I examine how presence and diegesis are approached and produced during the creative construction of the pro-filmic space[9] through the different eras of sound production in Indian cinema recognized worldwide, primarily for its voluminous film industry.

The Listening Field

As Susan Hayward has observed, the coming of sound in cinema introduced 'a crucial element to the registering of authentic reality'; henceforth, sound film 'was touted as being closer to reality' (2006, p. 359). Adding sound to the screen image could provide a perceptually realistic delineation of the site through the process in which, as film scholar Peter J. Bloom suggests, 'the experience of sound may become more spatially defined. By contrast with a two-dimensional image, the temporal nature of sound becomes related to the hearing subject's own location in any given space' (2014). The spatial characteristics of sound were recognized and explored after the coming of sound in cinema as an anchor to the story-world as the tangible setting of a site that could be associated through the spectator's lived experience of place.

Sound recording also opened up the palette for film practitioners to choose materials for providing spatial evidence in terms of site-specific details. In this palette, sonic elements such as voice, music, and effects, and background noises like natural and environmental sounds, along with other location-specific sounds that were recorded on site, primarily contributed to the realization of auditory settings since the time of the talkies. My research interest for this book lies in understanding the trajectories of this practice in India, that is, the way different layers of sound are recorded and spatially incorporated so as to produce sonic experience in film and audiovisual media in the contemporary digital realm. This trajectory perhaps poses a large historical canvas, but the singular focus of this project remains on a practice-led study of the creative practice of sound to construct and produce the spatiotemporal experiences in Indian film and audiovisual media through different technological phases. A qualitative evaluation of

9 The area in front of the camera's recording field is known as the 'pro-filmic space'.

the production practices and various methods used by sound practitioners working within the film industry, as well as description and analysis of a number of relevant works, will inform this inquiry. Despite the rather broad historical framework that this goal implies, my aim is to study film sound while questioning classical assumptions of the sound-image relationship in film and media studies and shift the focus onto the creatively constructed sonic spatiality in the diegetic world of film and audivisual media as a vital narrative strategy.

Given the sporadic scholarly attention to the field of sound, Indian films are generally criticized for producing transcendental sound experiences, exemplified by an overwhelming use of sloppy and poor sounding 'song and dance' sequences (Booth and Shope 2013; Morcom, 2016). In this practice, the careful incorporation and attentive organization of sounds are generally ignored in the narrative strategy of Indian film production (Rajadhyaksha, 2007; Gopalan, 2002; Ganti, 2012). There are indeed many examples from popular Indian films and media productions in which mindful sound design is not apparent; often a loud, high-pitched auditory setting is meant to provide a remote and imaginary sonic landscape. Challenging this popular notion, I intend to demonstrate that this generalized perception of sound practice in India is erroneous if we consider the historical trajectories of sound production. The advent of digital technology makes it possible to incorporate rich layers of sound in the production scheme of current types of Indian films and audiovisual media. A thorough study of this historical trajectory via detailed case studies and close readings of particular films beyond the popular and commercial so-called 'Bollywood' films – including regional films – focusing on the creative use of sound will help to put the inquiry, analysis, and observations in context. A practice-based perspective of attending to the methods and approaches undertaken by sound practitioners will provide the empirical evidence to qualify the claims; extensive interviews conducted with practitioners, based on critical self-reflection, are the primary sources of real-world knowledge in this book.

Following the trajectories of film and audiovisual media production in India, this study will reveal that sounds have been inconsistently rendered through various phases of production practices due to the effects of technological innovations and shifts. For instance, there have been phases, such as the entire period of the 'dubbing era' (1960s–1990s), that were less concerned about the quality of sounds, and gave more importance to the spectacular 'song and dance' sequences often placed within the flow of the narrative as interruptions and suspension of the story. However, there are also phases such as the 'digital era' (2001-present) where 'realistic' and more concrete representations of sound are observed, as mentioned earlier. The dominant factors that have determined these shifts are rooted in technological developments (Kassabian, 2013; Kerins, 2011, 2006; Lastra, 2000; Altman, 1992); certain aesthetic choices were made available by these technological applications, shaping the historical evolution of sound production practices in India. This history will be thoroughly studied in order to understand the various differing and concurrent practices, methodologies, and approaches, as well as the shifts in recording and organizing sound. The book will divide the trajectories of sound into its three most prominent markers, primarily based on the aesthetic shifts informed by technological changeovers and related innovations in sound recording. This historical division will be articulated in terms of three primary phases, namely: the optical or 'direct' recording era as it developed in the early talkies (1930s–1960s), the magnetic tape-based recording that triggered dubbing practices (1960s–1990s), and the contemporary digital multi-track sync recording and surround sound era (2001–present). Each of the subchapters in the book will take these three production phases as fundamental mechanisms and systems for certain kinds of aesthetics that emerge and permeate the sound production practices shaping film and media productions in India. Taking this model as a historical structure, the rest of the book will be devoted to a descriptive analysis of various modes and facets of sound production in India, namely Indian cinema and audiovisual media, from mainstream popular Hindi language films made in Mumbai to regional cinemas from Bengali, Malayalam, Tamil, among others, as well as a few audiovisual media works. This comprehensive study and detailing of sound production practices in India through a critical listening to a number of film and media examples will be contextualized in the light of film and media studies, film and media production, media art history, and sound studies.

In the arts and humanities, 'sound studies' has emerged and rapidly established itself as a vibrant and productive academic field resulting in the current profusion of scholarly writings in the broader areas of music history, musicology, performing arts, and culture focusing on sound and listening. Three consecutive compendia such as *The Routledge Sound Studies Reader* (2012), *The Oxford Handbook of Sound Studies* (2013) and *The Routledge Companion to Sound Studies* (2018) have been complemented with the *Journal of Sonic Studies* (Research Catalogue platform), and several other peer-reviewed journals that are entirely dedicated to the studies of sound. These publications show that the study of sound is indeed now a relevant area of research receiving wider academic attention within the broader contexts of music, performing arts, musicology, and music studies. Notwithstanding this rapidly growing body of work (Sterne, 2003, 2006, 2012; Born 2013; Théberge, Devine, and Everrett, 2015; Dyson, 2009, 2014; Demers, 2010; Novak and Sakakeeny, 2015; Blesser and Salter, 2009; Bijsterveld and Pinch, 2013; Bull et al, 2018), much of the attention has been invested in studying sound within an American and/or European film and media cultural context. Sound in other cultural contexts such as India has largely remained underexplored, even though India is one of the world's largest producers of film, music, and audiovisual media, with a formidable global presence as an emerging giant in the performing arts, music, and cultural production. This book intends to fill this void by developing a comprehensive understanding of the unique sound world of Indian film and audiovisual media production, through an examination of the historical developments of sound production practice and the corresponding aesthetic shifts in film and media.

A Sonic Framework

Film sound scholarship

The past few decades have seen the establishment of specialized scholarship for the study of film sound. The early writings of Rick Altman and Tom Levin on the history of sound technology have been instrumental in this developing field, examining sound's relationship to the moving image. Michel Chion's seminal works, including of course *Audio-Vision* (1994), have become canonical texts on film sound as part of a developing corpus of knowledge focused on understanding the audio-visual relationship operating on screen. These writings have consistently studied sound in relation to the visual image, whereby sound has been regarded as an element predominantly enhancing the cinematic narrative, a reinforcement of the visual testimony on screen, and/or underlining the emotive potential of a scene. This scholarly stance considers the 'soundtrack' as a secondary and one-dimensional accompaniment to the visual storytelling without enacting an autonomous impact on the multi-sensory cinematic experience. Chion argues that we don't see images and hear sounds as separate channels, we audio-view a trans-sensory whole (1994), but in multi-channel surround environments, some off-screen sound elements indeed demand having autonomous impacts beyond the viewing aspects of cinematic experience. As charted in my book *The Auditory Setting* (2021a), in recent works, however, an apparent shift seems to be emerging. Among contemporary scholars, Giorgio Biancorosso (2009), Randolph Jordan (2012) and myself (2016, 2017,

2018, 2021a, 2021b) question the overarching comparisons of sound with image and emphasizing the role of sound and the potential of listening by taking sound as a specific and separate area of research. The idea that sound can be studied separately from the moving image, and that sound in movies is more than the flat and screen-centric soundtrack following the visual narrative, has destabilized the notion of a 'soundtrack'. This term has retained its historical relationship to an optical track on the filmstrip, mixed with music, thus retaining a sense of linearity and one-dimensionality literally alongside the visual narrative. Following Biancorosso and Jordan, and continuing my own scholarly work (2021a, 2021b), in this book I will question the traditional idea of the 'soundtrack' in Indian cinema, advocating for a more fluid and malleable multilayered sonic environment. This is particularly true for the production methodologies evident in digital production, due to the fact that the effects of spatial practices with sound transcend a linear representation of the soundtrack towards a spatially evocative sound environment that creates elaborate and fluid cinematic spaces where the epistemological grounding of the sounds (Branigan, 1989) to their respective screen-centric visual sources is reordered. The wider off-screen diegetic space available in digital cinema systems (Kerins, 2011), appearing to be immersive and enveloping toward the embodied experience of the listening audience (Ihde, 2007), also opens up the scope for multiple interpretations of sound (Nancy, 2007; Chattopadhyay, 2014). This new setting, as I pointed out in my earlier research (2021a), leads to an interactive and flexible sonic space (Dyson, 2009) that often appears to render itself beyond the visual image and constraints of the screen. Reading the actual impacts and manifestations of creative sound practice and implications through various phases of production becomes relevant and necessary in this altering discourse in film sound research.

There have been sporadic efforts and attempts made to study the use of sound in Indian cinema, but much of this scholarship has been focused on the use of songs (Mukherjee 2007, et al.), voice, and background musical scores (Booth and Shope, 2013; Morcom, 2016) in popular Indian films. Indian film music has indeed been a focused area of scholarship; for example, the work of Alison Arnold (1991) analysed Hindi film songs along with the studio practices that produced these songs. Music critics like Bhaskar Chandravarkar and Ashok da Ranade have also illuminated the field, writing essays about music in Indian films. Indian filmmakers like Satyajit Ray (1976, 2011), Ritwik Ghatak, Mani Kaul, and Kumar Shahani have commented on the use of music in Indian film, among which Kaul's writings (1977, 1983, 1991) have been noteworthy for their erudition, insights, and incisive understanding of classical Indian aesthetics around visual and aural perspectives and temporalities.

The history and proliferation of sound technologies such as the radio and the gramophone in the 1920s and 1930s India are researched in the ethnographic writings of Stephen Hughes. Early cinema and sound in the talkies have also found sustained scholarly attention. The edited volume *Aural Films, Oral Cultures: Essays on Cinema* (2012) probes into Indian films from the early sound era. The essays included in this book contain important documentation of the period of the so-called 'talkies' from the early sound era. These historical engagements with sound in early Indian films bring together debates on the transmission of sound technologies from the West and their applications in India. Focusing on one of the early talkie studios, New Theatres, this study actively engages with the proliferation of sound technology in the 1930s from the early direct sound era in Indian cinema and the socio-political contexts within which such technological changes occurred. Musicologist Gregory Booth's writing (2013) also studies the circulation of sound technologies such as the gramophone, radio, and cassettes on the practice of musical composition and orchestration in the early Hindi film industry.

As I have mentioned, the edition of the *Journal of the Moving Image* (2007) from Jadavpur University's Film Studies Department focused entirely on the aesthetics of sound in Indian cinema. This issue, where I had my first article published, was a crucial intervention in this field. The issue indeed catapulted a critical interest in film sound research in India and, on a very personal level, launched my research into sound.

These infrequent scholarly interventions are often focused on the contextual knowledge around film (sound) history, such as the socio-economic conditions that shaped the technological transformations in early talkies in India and studio-era sound practices. While this social scientific approach is revealing from an interest in film history and theory, social, and cultural studies, it is, however, not adequate to grasp the vast and varied fields of sound in Indian film and audiovisual media from aesthetic, historical, and production perspectives. A creative practice with sound, particularly components like ambience and sound effects, which are underlined by sound practitioners as critical elements in film sound organization (Chattopadhyay, 2021b), is still underexplored and requires further study. I intend to address this lack by developing a systematic, creative practice-based understanding of the unique sound world of Indian film and media. At the core of this research is a consideration for a phenomenological approach of attentive listening to the sound experience in Indian films and audiovisual media, as well as self-reflective analysis of this personal experience and critical engagement with the sound practitioners themselves in extensive fieldwork as primary sources of knowledge, rather than relying on contextual information to write a history.

The ever-growing field of Sound Studies across the globe

In the growing field of sound studies, in-depth as well as inspired and dedicated inquiries into the sound practice in the Indian context should be accommodated and appropriated, contributing to the field of sonic research. Jay Beck and Tony Grajeda quote Rick Altman in the introductory chapter 'The Future of Film Sound Studies' in their edited anthology *Lowering the Boom: Critical Studies in Film Sound* (2008) to emphasize the appropriateness and legitimacy of specialized investigations into film sound studies and studies of sound (media) production as part of the growing field of sound studies. Within an emerging discourse of practice-based research in film and media studies, the in-depth study of sound practice in India demands to be pursued from the novel perspectives of sound studies – a perspective that allows for studying sound beyond the domination of the visual image and beyond the constraint of the screen, allowing the research to focus on the emerging spatiality evident in Indian sound practices, especially in light of the digital innovations facilitating spatial practices in contemporary film and media arts, including India. Therefore, this inquiry will be primarily informed by studies in sound production (Holman, 1997, 2002; Kerins, 2011; Sergi, 2004; Sonnenschein, 2001; Buhler, Neumeyer, and Deemer, 2010; LoBrutto, 1994) and drawing upon Sound Studies (Sterne, 2012; Born, 2013; Théberge, Devine, and Everrett, 2015; Dyson, 2009, 2014; Demers, 2010; Novak and Sakakeeny, 2015; Blesser and Salter, 2009), with investigations of diegesis, mimesis, and narration in cinema and other media art (Bordwell and Thompson, 1997; Percheron, 1980; Burch, 1985, 1982; Kassabian, 2013; Weiss, 2011), the notion of sonic presence (Doane, 1985; Skalski and Whitbred, 2010; Grimshaw, 2011; Reiter, 2011; Lombard and Ditton, 2006), and the history of Indian cinema (Rajadhyaksha, 2007; Gopalan, 2002; Mukherjee, 2007; Allen, 2009; Carrigy, 2009a, 2009b; Chattopadhyay, 2015, 2016, 2017; Sengupta, 2007). The research will draw from studies in film sound (Altman, 1992; Balázs, 1985; Biancorosso, 2009; Chion, 1994; Lastra, 2000; Bloom, 2014), sound perception and cognition (Bordwell, 2009; Waller and Nadel, 2013), phenomenology of sound (Ihde, 2007; Sobchak, 1991; Nancy, 2007). This book will have good company among a growing body of sound studies literature across the globe beyond its Eurocentric delimitation: for example, *Hungry*

Listening: Resonant Theory for Indigenous Sound Studies (Robinson, 2020) and the edited volume *Remapping Sound Studies* (2019).

Drawing inspiration from these newer works, this book will examine the aesthetic shifts in sound practice in the light of technological innovations and the historical evolution of film sound in India to underscore complex interplays between the Western hegemony of technology pressuring to embrace modernity and the emergence of a syncretic, hybrid, and assimilated form with traditionalist traces in Indian film sound that negotiates Indian epistemology with modernist pressures. The critical engagement with relevant literature in the non-Western canons will inform the research and will be considered useful for the generation of new knowledge in the ever-growing field of sound studies beyond its inherent Eurocentric vision. By drawing attention to an ignored site of inquiry, that is, sound practices in India, this book will help address a number of critical issues in contemporary sound studies, such as decoloniality, Global Souths, and non-Western epistemologies, as well as fundamental topics such as storytelling, diegesis, and realism.

Technological innovations towards the digital realm beyond Eurocentric sound

Film scholars have already pointed out that technical advances in cinema have affected the spatial aspects of sound creation in the various eras of sound technology (Bordwell and Thompson, 1985). I assume the evolving sense of spatiality in Indian cinema is primarily supplied by additional layers of environmental sounds while other layers such as voice, music, and sound effects remain largely screen-centric through technological transitions from monaural to stereophonic and digital surround environments. There are instances of the voice carrying spatial information, particularly in the case of digital multi-track synchronized or 'sync' sound practice, as Mark Kerins shows (2011). Instead of dubbing inside a studio – a practice that was standardized in many national cinemas similar to India's, for a large part of their histories (1950s–2000) – the sync track would include some spatial information directly recorded from the location such as original reflections from indoor rooms ('room tone') and so forth. This would add to the overall sense of spatiality, but it is the layer of environmental sound that would carry the *primary* information as well as being the basic tool in the hands of the sound practitioner to reconstruct the site and to enhance the 'ultrafield' of the sonic experience (Holman, 2001; Kerins, 2011). Following this argument, this book considers all the layers of sound horizontally, discussing voice and music alongside environmental and incidental sounds and sound effects in its research, critical analysis, and commentary.

In my previous book *The Auditory Setting* (2021a), I focused on a critical study of environmental sound use in global cinema. There, I noted that historically, sole reliance on the voice to carry out the primary threads of narrative has been lessening ever since other channels of sound have opened up to distribute the screen-centric appeal of films towards an expanded cinemascope and surround sound (Rogers 2013). This development can be seen as a shift away from the 'vococentricity'[10] of cinema (Chion 1994) towards a more inclusive realm of digital systems. Since the 1990s, a large-scale conversion from analogue recording, analogue production practices, and optical film exhibition to digital technologies has taken place. Digital technology has been integrated into the production and post-production stages of filmmaking. The ramifications of these developments have been far-reaching – it was particularly evident in the way cinematic experience changed through the radical use of sound, such as multi-track synchronized sound recording and surround sound design. In very recent times, digital multi-channel surround sound systems like Dolby Atmos and Auro-3D have altered the way in which the film 'soundtrack' is rendered and reorganized in cinema. These newer environments have reconfigured the spectator's experience of the cinematic space, contrasting considerably with earlier predominantly screen-centric mono and stereophonic settings, integrating and augmenting the monaural and stereophonic aesthetics into the surround environment through the reordering of the spatial organization of cinematic sound. Following these transitions, contemporary Indian cinema facilitates specific practices of sound to create cinematic experiences which, I show in this book, are spatially wider, more elaborate and fluid compared to the screen-centric mono-aural soundtrack or the flat surface of a slightly wider yet still within the screen, stereophonic composite soundtrack. These earlier organizations of sound in Indian cinema anchored the story-world narrated on screen, altogether ignoring the specifics of site in diegesis, yet evoking sound's emotive potential by using post-synchronized effects and background music. I will argue that digital era's sound practices incorporate the surround multi-channel design of site-specific and other locative sounds that handle the site in spatially

10 'Vococentricity' is the term introduced by film theorist Michel Chion that refers to the tendency to organize the soundtrack in a hierarchy with the voice as the most important element. The other components of the soundtrack (e.g., music, Foley, ambience) are organized in relation to their interaction with the human voice (dialogue).

perceptible ways, rather than catering to a screen-oriented audio-visual contract as Michel Chion postulated earlier.

Chion's views of film sound (1994, 2009) may appear questionable in the current analysis of sound practice that enables more layers of sound to be included in the spatial organization of sound in cinema. In *The Auditory Setting* (2021a), I have elaborately discussed how, contrary to what Chion had envisioned (although primarily discussing American and European films), today's expanded universe of a (post-)global film sound experience builds multi-channel environments, which assume that audiences will understand sounds originating in the surround channels to be part of the same diegetic space as those originating on screen (Kerins, 2011) expanded with elaborate and autonomous details beyond the screen (e.g., the sound of an overhead airplane or a flying bird behind our head) in the Dolby Atmos sound environment of a cinema theatre. Innovative practices with sound in the digital era have led to newer experiences of cinema in which spectators engage with cinematic sites through immersive listening, employing spatial cognition of sound. These practices need new theoretical models and approaches in film sound studies beyond Chion and his contemporaries. The new approach will need to cater to the studies of sound in the contemporary digital milieu overcoming domination of the moving image with screen-centric interpretive tendencies. The approach will also need to include the various innovative threads of sound practices in discourse beyond the prevailing American and Eurocentric outlook of film sound studies. In this context, the current book project enables a refreshing, if not innovative, scholarly framework to study the increasingly important layer of sound in the digital realm, namely ambient sounds, and their role to enrich the cinematic experience by exploring sound's spatial dimensions. The 'spatial turn' (Eisenberg 2015) in Indian film sound production manifested by the use of site-specific sounds will be studied in the light of its trajectory of development from monaural synchronized sound recording and post-synchronous dubbing to the contemporary digital multi-track synchronized or sync recording, as well as from monaural and stereophonic mixing to digital surround sound design. Challenging prevailing American and Eurocentric sound scholarship, a systematic study of the use of sound in Indian cinema is considered not outside of this larger theoretical corpus but as a useful intervention and scholarly addition. It is needless to say that some theoretical references for this book were made having mostly examples of American and European films as case studies (e.g., Holman, 1997, 2002; Kerins, 2011; Sergi, 2004; Sonnenschein, 2001; Buhler, Neumeyer, and Deemer, 2010; LoBrutto, 1994, Sterne, 2012; Born, 2013; Théberge, Devine, and Everrett, 2015; Dyson, 2009,

2014; Novak and Sakakeeny, 2015). Nevertheless, these references create the project's epistemological grounding of film sound studies where examples from Indian films can be historicized, theorized, and contextualized to cultivate an Indian epistemology in the field of sound studies as a novel and much-needed contribution.

Further Theoretical Considerations

Audio-audio relationship

Studying sound in film and media has all too often stagnated in investigating the audio-visual relationship (e.g., Chion, 1994). It would be worthwhile to shift the perspective on the complex relationship between different sound components: namely, voice, music, sound effects, and ambience or environmental sounds. There are typical hierarchies between these components, and often one component (e.g., voice) is given more importance over others (e.g., ambience) in narrating a story or depicting a situation.

In *The Auditory Setting* (2021a), I have discussed these hierarchical relationships, namely the voice as a primary sound component that includes dialogue between characters carrying the primary information of the narrative communicated to the audience (Bordwell, 1997). Amy Lawrence argues that in narrative cinema 'the synchronization of image and voice is sacrosanct' (1992, p. 179) emphasizing the necessity of the stricter method of sound production in regards to the voice, which must be connected with a 'body' on the screen. Mary Ann Doane affirms that dialogue or the use of the voice 'engenders a network of metaphors whose nodal point appears to be the body' (1985, p. 162). She further states that the sound of the character's speech or voice is strictly 'married to the image' (1985, p. 163) on the screen, making it creatively rigid as a sound component in cinema for spatial maneuvers unlike background sounds. Voice attributes in cinema are less 'spatial' in nature. Moreover, the post-synchronized voice, as produced by dubbing or similar practices, is 'disengaged from its "proper" space (the space conveyed by the visual image) and the credibility of that voice depends upon the technician's ability to return it to the site of its origin' (Doane 1985, p. 164), a condition which can be achieved by the technician's creative and innovative use of ambient sound components.

As indicated in *The Auditory Setting* (2021a), non-diegetic music in films creates situational feelings and underscores certain emotions. Film music is used 'largely to set mood or elicit a particular emotional response from the audience' (Kuhn and Westwell, 2014). Film music tracks and sound effect tracks 'establish a particular mood' (Doane, 1985, p. 55) instead of providing a sense of space. Sound effects are also important for the narration, and for creating feelings of tension and horror. In the mixing stages, the hierarchy of different sound components follows certain conventions: 'Sound effects and music are subservient to dialogue and it is, above all, the intelligibility of the dialogue which is at stake, together with its nuances of tone' (Doane, 1985, p. 55). In this hierarchy, ambient sound remains fluid and malleable in the hand of the practitioner.

While elaborating the processes of sound production in Indian cinema, the practitioners interviewed in my book *Between the Headphones: Listening to the Practitioner* (2021b) underline that ambient sounds provide a specific atmosphere of a site in the construction of the pro-filmic space. To the sound practitioners of Indian cinema,[11] ambient sound injects life and substance not only to what we see on screen but also to the off-screen diegetic world. The practitioners use layers of ambient or environmental sound to construct the experience of reality by artistic means. These sounds help mount the sense of a specific place in Indian films consolidating their spatial dynamics.

Film space

The term 'film space' is defined as the space that the spectator encounters – a space that is organized through time (e.g., the linking of shots through sound editing). On the other hand, the area in front of the camera's recording field is known as the 'pro-filmic space' in the glossary of film terms (Sorfa, 2014). Combining these two definitions, I argue that the choice and arrangement of pro-filmic space substantially affect the spatial dynamics of the mise-en-scène of sound or 'mise-en-sonore' (I have used this loose coinage in my writings: Chattopadhyay 2016, 2021a) equivalent to the *auditory setting* – the actual sonorous environment that appears to the experience of the auditor – a setting that in turn influences the verisimilitude or believability of a film in the ears of the audience member. Film scholars Annette Kuhn and Guy Westwell define pro-filmic space as:

> The slice of the world in front of the film camera; including protagonists and their actions, lighting, sets, props and costumes, as well as the setting

11 See: interviews with prominent Indian sound practitioners such as Dipankar Chaki and Dileep Subramanium speaking about the ways they use ambient sounds in Indian cinema in *Between the Headphones: Listening to the Sound Practitioner* (Chattopadhyay, 2021b).

> itself, as opposed to what eventually appears on the cinema screen. In studio-made fiction films, the profilmic event is a set constructed for the purpose of being filmed. At the other extreme, in observational documentary forms like direct cinema, filmmakers seek, as a fundamental element of their practice, to preserve the integrity of the real-life space and time of the profilmic event. Many films occupy a middle ground in their organization of, or relationship with, the profilmic event: as for example in the case of location-shot. (2014)

As the narrative rendering of the pro-filmic space in sonic terms, 'mise-en-sonore' or auditory setting[12] is understood as the simulated space of the fictional site, which is constructed in cinematic experience using environmental sound. This playful formulation of the term expands on James Lastra's conceptualization of the term 'pro-filmic space' (2000) and draws on the definition of the same term by Annette Kuhn and Guy Westwell to explain roles of environmental sound in the process of narrative storytelling 'to preserve the integrity of the real-life space' (2012, p. 333). These ideas are discussed in detail in *The Auditory Setting* (2021a).

(Quasi-)diegesis and mimesis in film sound

It is necessary to consider the state of how the story-world of Indian cinema is formed given the rich storytelling traditions existing in the subcontinent ranging from oral storytelling to the audiovisual such as *Patachitra* and *Pater Gaan*.[13] There is a limited vocabulary to discuss such syncretic practices in the Western film sound canon. The Ancient Greek notion *diegesis* denotes a process to narrate a contained story-world, which, as Mary Ann Doane has noted, is the internal space of the cinematic universe framed and constructed by the technical tools of filmmaking (Doane, 1985). As shown in *The Auditory Setting* (2021a), translated to sound, this term could be understood as relating to the creative layers of sounds that are made to emanate from the story space in which events occur. Claudia Gorbman defines diegesis as 'the narratively implied spatiotemporal world of the actions and characters' (1987, p. 21). Both Doane and Gorbman use the terms 'space' and 'world',

12 This term is formulated as a useful coinage to study the spatial practice of film sound in *The Auditory Setting: Environmental Sounds in Film and Media Arts* (Chattopadhyay, 2021a).

13 *Patachitras* are visual storytelling styles in Eastern India with accompanying songs, *Pater Gaan*, weaving stories together as a performative element. The songs follow traditional or modern tunes with old or new words composed for the specific stories told. The song sets the story along with the visual scroll and brings it to life in performance.

underscoring the constructed nature of the film space during narrative storytelling. In *The Auditory Setting* (2021a), I have discussed how the term 'diegetic sound' helps to consider the sounds that inhabit the constructed world and whose sources are visibly present within the space of filmic events. Likewise, any sound outside of the story-world of a filmic universe is termed non-diegetic. The most common example of non-diegetic sound is the amply used sound component in Indian cinema: BGM or background music, which the characters do not hear, and which is not present in the space of filmic events; therefore, BGM is not grounded in the filmic reality apart from expanding certain emotional situations to manipulate audience responses. While songs sung by the characters are diegetic, many of the accompanying instruments in the song remain invisible, and therefore non-diegetic. The stereotypical song and dance sequences in Indian cinema are therefore a hybrid of diegetic and non-diegetic elements – which I term 'quasi-diegetic' – indicating the illusion of diegesis in sound use. Site-specific sound effects, reflected voices, incidental, and ambient sounds can be regarded as the means of reinforcing a sense of diegetic realism by enhancing a site's believability in the story-world's spatial environment, to make the sound and image (of the site) 'credible' (Wayne, 1997, p. 176). Examining the spatial practice of sound to construct the site's relative presence in the story-world must take into account and interprete the concept of diegesis in the context of Indian films.

As I noted in *The Auditory Setting* (2021a), historically, diegesis is understood as the process of illustrating the story-world with all the narrative elements that are shown or inferred within the filmic content. The process allows for a certain mediated discernment of the phenomenal world within the story, including all the physical pro-filmic spaces framed inside the film, be they indoor or outdoor locations, or film sets and studios. These spaces might be narrated with their auditory features, characteristics, atmospheres, and 'soundmarks'[14] (Schafer, 1994) in order to establish their presence in the mind of the audience, who can construe a diegetic world from the recorded and (re)presented sonic materials as they take in tiny aural hints to interpret contours of the sites from the relative volumes and spatial matrixing of these sounds. Noël Burch (1980) states that diegesis includes a description of the narrative action proper, including places, people, clothing, and sounds. In order to build a mediated story-world,

14 Inspired by the word 'landmark', the term 'soundmark' was coined by Canadian composer R. Murray Schafer to refer to a site-specific sound that is unique or possesses locational qualities, making it noteworthy in its usage to represent a place sonically.

diegesis is a sound production methodology whereby sound components like the voice and effects take crucial roles in telling a story. Film scholar Edward Branigan suggests that diegesis is the 'implied spatial…system of a character – a collection of sense data which is represented as being at least potentially accessible to a character' (Branigan, 1992, p. 35). Both scholars point out the necessity of the spatial component in narrative (re)construction of the pro-filmic space within film space. As we will see, sound practice in the digital era of Indian cinema brings such spatial components to the fore.

It is not new that the shifting sound practices in American and European cinema, impacting diegesis, have been studied using the theoretical corpus of narration and storytelling. However, in Indian film scholarship, this field of sound studies related to narration is relatively understudied. Writing on the specifics of narration in the digital realm of American cinema when using DSS (Digital sound systems), American film sound scholar Mark Kerins argues that 'filmmakers have…relied on ambient sound in the "surrounds" to set up diegetic spaces, and this trend has certainly continued with movies employing DSS' (Kerins, 2006, p. 44) – hinting at what I aim to underline in this book about an emergent diegetic spatiality and sonic realism found in the digital era of Indian indie cinema. In a later chapter of this book, I will explore sound practice in the Indian indies (independent films made outside the studio system, especially with regard to funding, working outside the filmstar system). The sonic realism approached in this breed of new Indian films is studied with careful attention so as to underline the fact that these emerging films are far more open to experimentation with the creative potential of sound.

In the above statement, Kerins writes about the narrative strategy of the practitioner embracing digital systems to produce enhanced diegetic spatiality, which contributes to the sense of sonic realism and presence. As he writes, the apparent completeness of the constructed aural environment in cinema suggests the relatively high (or, arguably, the highest) degree of presence in the rendering of the film space as (re)presented in the story-world. This story-world would be considered diegetic if the elements that belong to the film's narrative universe are included in the storytelling. In Indian cinema's analogue-era, diegesis is a crucial narrative device, as the dialogue often seems to 'tell' the story as narrators rather than using other mimetic elements of 'showing' such as ambient sounds. This book shows that such normative structure is destabilized after the digital.

The notion of diegesis in relation to sound has an aspect of mediation embedded within. This mediation allows the narrator to control what is revealed in the narration: objects, situations, spaces, and characters that inhabit the story-world. The process of mediation follows two stages of sound production: the recording of sounds from the pro-filmic space, and the sound design in the studio to recreate and produce the sonic film space that I term the 'auditory setting'. Taking a parochial, and ideological stance (Kelman, 2010) against the low-fidelity environmental sounds and the mediation of sounds recorded from situated environments, Schafer stated: '[W]e have split the sound from the maker of the sound' (1994, p. 90). In this book, I will show how these 'torn' and 'ruptured' sound recordings go through further mediation in the hands of sound practitioners whereby sonic materials are processed and organized spatially to create new diegetic worlds that are often abstracted from their sites of recording – the realistic pro-filmic spaces of the locations or sets. This is particularly heard in commercial Indian cinema due to its generally escapist approach and tone often manifested in the heavy noise cleaning, rough editing, and compression of sounds.

Since film is always framed by the camera (and sound recorder), early cinema scholars have argued that 'it is therefore a diegetic form and not a mimetic one' (Prince, as cited in Kassabian, 2013a). Opposed to the basic principles of diegesis (i.e., narration and depiction), 'mimesis' suggests imitation (Weiss, 2011; Dumouchel, 2015). In essence, while diegesis 'narrates' the action, mimesis 'shows' the action (Kassabian, 2013a). I have discussed in *The Auditory Setting* (2021a) how the merging of these two approaches takes place in cinematic narration. Likewise, film theorist Anahid Kassabian suggests that narration through sound in cinema combines both strategies to a certain degree: 'Surely all realist film forms are both diegetic and mimetic in significant proportions, and it might be more interesting to consider how, when, and why those proportions shift in one direction or another' (2013a, n.p.). In this book, I intend to examine if these proportions shift in the context of the move from analogue to digital production frameworks in the context of Indian cinema that often conventionally depends on 'telling' rather than 'showing' advocating for a more complex diegetic model of storytelling. However, there is a recent shift. I will argue that with contemporary digital sound techniques, the pro-filmic spaces in Indian films appear more present, being spatially wider, more elaborate and fluid compared to the screen-centric monaural soundtrack or the flat surface of the stereophonic composite soundtrack. The digital realm's sound practice has incorporated the multi-track synchronized sound recording and surround spatialization of sounds in a more mimetic process of representation, showing locations with intricate details, instead of employing other overly controlled ways of narrating, as if 'holding a mirror to…nature' (Dumouchel, 2015, p. 51), particularly in indie films. It is no surprise that the sites appear mimetically more present in the digital era in Indian cinema than in previous eras of sound production, as it has become possible to render 'sounds with an increased exactness' (Beck, 2008, p. 72). In this book, I locate a shift occurring in Indian cinema from the diegetic to the mimetic as we move from the analogue to the digital, producing an intensified sensation of realism.

Sonic realism, presence, auditory spatiality, and situated-ness

The idea of sonic realism is very much connected to the notion of 'presence', which is understood as the degree to which a medium can generate a seemingly accurate reproduction of objects, events, and space – representations that look, sound, and/or feel perceptually real. Mary Ann Doane claimed that 'concomitant with the demand for a lifelike representation is the desire for "presence", a concept which is not specific to the cinematic soundtrack but acts as a standard to measure quality in the sound recording industry as a whole. The term "presence" offers a certain legitimacy to the wish for pure reproduction' (1985, p. 163). The word 'pure' as Doane uses it denotes a desire for a natural, extant, and genuine registering of the sound in the recording. Doane made this reflection on presence before the arrival of digital technology in sound recording, a new realm that not only complicates but also supports her statement. I have pointed out (2021a) that her conceptualization of presence was based on analogue optical or magnetic sound recording practices. In the digital realm, presence gains currency in the digital sound system's capacity to produce the 'complete sonic environment' (Kerins, 2006), presenting more detailed sonic information from the location and surrounding the audience member in a spatially richer perspective in terms of frequency and dynamic ranges – as evident in contemporary Indian films made since digital innovations from 2000s (e.g., films made for multiplexes or OTT platforms).[15] These films tend to create an 'immersion in the filmic environment – audiences are, ...aurally, literally placed in the middle of the action' (Kerins, 2006, p. 44). In current sound research, presence is defined as the 'feeling of being present' (Reiter, 2011, p. 174), as the 'perceptual illusion of nonmediation' (Lombard and Ditton, 2006, p. 9), even in studio-constructed experience. These interpretations of presence suggest a tendency towards the mimetic representation of sound in film. In the digital production of the contemporary cinematic environment in India, the pro-filmic space appears

15 An OTT (over-the-top) platform is a Web-based media service offered directly to viewers via the Internet, such as Netflix, Hulu, Amazon Prime Video, bypassing the conventional cable network, broadcast systems, and satellite television channels.

with an intensified sense of sonic realism particularly due to the inclination to employ digital sync sound techniques and methodologies. As I noted in *The Auditory Setting* (2021a), this notion of presence helps to explain how audiences experience a feeling of 'being there' bodily, or 'participating in' rather than 'observing from outside'. In this book, I will demonstrate how an aesthetic of 'spatial fidelity' (Kerins, 2011) pervades in the digital realm. This notion of spatial fidelity operates closer to 'spatial realism' (Altman, 1994) and aural situatedness as opposed to the term 'sound fidelity' that Jonathan Sterne discusses in *The Audible Past* (2003).[16] In Indian cinema, the experience of aural situatedness is re-emergent as we read the trajectories from the earlier dubbing era to the digital era. However, this book will locate how such a sense of aural situatedness was also found in the talkies of the direct sound era when all dialogue, songs, ambience, and incidental effects were recorded directly on the optical film during the performance, roughly in the 1930s–1940s. Drawing attention to the spatial fidelity of the recording capabilities inherent in the digital systems – as do Kerins, Sergi, and Holman – I emphasize the spatial faithfulness and lifelikeness that digital sound production provides in Indian film and media works.[17]

This book asks the question whether all the subtler aspects of the acoustic worlds of the urban and rural sites of India are narrated truthfully and mimetically even within the flexible and inclusive palette of digital sound production in Indian cinema. In most cases, the noisy aspects of the sound recordings are often controlled and sanitized by editing and advanced noise

16 As Sterne suggests, sound fidelity is a social construct and a social choice. The preference for a 'better' sound here hinges on the predominantly social and cultural aspects of sound reproduction. He questions the general assumptions about digital audio to be lacking in life and naturalness, arguing that 'digital recordings have as legitimate a claim on sonic experience as their analogue counterparts' (Sterne, 2006). The sound practitioners I interviewed for my research seem not to agree: while analogue sound is seen as warm and human-like, digital audio, in their opinion, is cold and clinical. Their opinion can be based on a social sense of nostalgia and loss due to the large-scale and revolutionary conversion of the filmmaking process in India from analogue to digital since the 2000s. A more detailed explanation of the term fidelity can be found in *The Auditory Setting* (2021a).

17 Later in the book, I will show and discuss in detail that digital sound recording and production frameworks in India have introduced a number of creative possibilities, including a significantly larger dynamic range, which is a fourfold improvement over the monophonic and almost double that of the stereophonic format; a larger headroom – a major improvement over both the monophonic and stereophonic formats; discreet channels for multi-channel systems, like Atmos; wider panning for sound spatialization; and full-frequency channels with a consistently flatter response than any analogue counterparts. These technical improvements, I argue, contribute to faithful recordings made on location and faithful-to-original sound design deployed in the studio. These claims are substantiated by discussion with Indian sound technicians.

reduction techniques to provide cleaner sounds.[18] The typically syncretic, chaotic, and inchoate structures of Indian cities are reflected in the multiple layers of sounds from pre-industrial, industrial, and post-industrial eras, simultaneously active in juxtapositions or in contrapuntal relationships with one another. The urban sound environment in most Indian metros is therefore overly noisy, and sonically overwhelming and disorienting for the listening subject (Chattopadhyay, 2014, 2020, 2021a). This amount of noise is not always heard in the augmented sonic environment of Indian films, where more sanitized and gentrified accounts of these sites are presented. This spatial atmosphere can no longer be understood as a linear and one-dimensional 'soundtrack', rather it might be called a 'sonic environment' or an evolving 'cinematic soundscape'. I have discussed in *The Auditory Setting* (2021a) that the notion of soundscape as postulated by Murray Schafer aims 'to draw attention to imbalances which may have unhealthy or inimical effects' (1994, p. 271). This 'moralizing' (Kelman, 2010; LaBelle, 2006) tendency applied to controlling the incoming ambience by means of 'acoustic design' strongly corresponds with the 'sound design' deployed in cinema in general, even in contemporary Indian cinema of the digital era, involving editing and advanced noise reduction. The underlying intention is to transform 'lo-fi' sounds into 'hi-fi' sounds, removing 'noise' contents while prioritizing the potential entertainment and enjoyment of audience members. According to Schafer, 'lo-fi' sounds are 'overcrowded, resulting in masking or lack of clarity' (1994, p. 272); they have a lower signal-to-noise ratio and tend to impose 'an increased level of disturbance upon the body, society and the environment' (LaBelle, 2006, p. 202). This compulsion of achieving clarity in the cinematic soundscape trading off mimetic truthfulness leads the Indian sound practitioner to often use sporadic 'soundmarks'[19] instead of accurately capturing the sounds present at the cinematic location, as I will argue in this book. This tendency to underline a particular sound, often at the expense of many other site-specific sound elements emanating from a specific site, intends to sonically compensate for the noise reduction and the editing of many sync sound layers in post-production.[20] These film

18 For an elaboration of this statement, please refer to the interviews with Indian sound editors and sound designers in my book *Between the Headphones* (2021b).

19 According to Schafer, a soundmark is 'a community sound which is unique or possesses qualities which make it specially regarded or noticed by the people of that community' (1994, p. 10).

20 'Post-production' is work done on a film (or recording) after filming or recording has taken place. In post-production stages, the editing, processing, designing, spatial organization, and mixing of sound are performed.

industry norms, practical rules, and film production regulations embedded in the essentially functional aspects of film sound creation in India often tend to thwart the artistic potential of the sound practitioner, keeping the practice away from enriching the film's fuller auditory spatial experience.

Sound Methodologies

Historical outlining and locating sonic transitions

In three respective subchapters of the book's second chapter, I will examine the nature of sound usage in various corresponding and intercepting phases of sound production in India: 1) direct optical recording (monaural), synchronized sound; 2) analogue magnetic recording, dubbing, monaural and stereophonic mixing; 3) digital sync recording, surround design and mixing. Critical listening and reflective analyses of film and audiovisual media in the later chapters will follow this historical outlining.

I have already observed that sound has been inconsistently rendered and produced through various phases of Indian production practices. I mentioned earlier that there have been phases of sound practice in Indian cinema, such as the entire 'dubbing era' (1960s–1990s) that cared little about the site, giving more importance to typical 'song and dance' sequences. However, there are also phases such as the 'digital era' (2001–) where the realistic and concrete representation of site is observed. The dominant factors that have determined these shifts are assumed to be rooted in technological developments in film sound as film scholars point out in the context of American cinema (Kassabian, 2013a; Kerins, 2011; Lastra, 2000; Altman, 1980). These perspectives are valid for Indian cinema too. Certain aesthetic choices were made available by these technological changes shaping up the historical evolution of sound practice in Indian cinema. Questions can be raised about whether cinema technology influences aesthetics in filmmaking and whether this assumption is technologically deterministic. In defence, I argue that technological innovations do not necessarily determine production aesthetics, but they facilitate certain options and aesthetic choices that the practitioners choose as their tools and methods. This argument is substantiated by Mark Kerins in his book *Beyond Dolby*, where he claims that film history is rich with examples of 'technology influencing aesthetics' (Kerins, 2011, p 54). For example, the emergence of sound, colour, and magnetic tape initiated deep changes in corresponding aesthetic features in cinema; magnetic recording paved the way for the asynchronous mode of sound reconstruction in India cinema (e.g., dubbing and ADR). As I have noted (2021a), film scholar Rick Altman also articulated the implications of sound technology on cinema in his seminal writings (1980, 1992) arguing that sound technologies shaped certain aesthetic practices in Hollywood.

For the purposes of historicization, the book divides the trajectories of sound practice in India into three prominent markers primarily based on aesthetic shifts informed by technological innovations and related changeovers. I have assumed that the digital sound system produces a different experience than that of the earlier generation of mono or stereophonic environments with dubbing and Foley replacing the original recordings made on location. These earlier processes of post-synchronization, heavily compromised by the spatial characteristics of sounds, were perceived less dynamically and with relatively less depth, and the delicate relationship between sound and space in Indian cinema varied substantially from one setting to another. This trajectory of the use of sound in Indian film and audiovisual media works needs to be studied historically to understand how sound adds to the sense of verisimilitude and presence while constructing the film space and other mediated environments. Therefore, I divide sound production practices into three primary phases: namely, optical monaural direct recording era (1930s–1950s); the magnetic recording and dubbing era (1960s–1990s); and contemporary digital multi-track location sync sound and the surround design era (2001–present). Questions can be raised as to whether these three historical markers as a loose tri-partite model are historically accurate, or if perhaps they limit the scope of the research. I would argue that the loosely based three-part division is a heuristic – a pragmatic and useful portal to read the fragmentary, uneven and often intercepting trajectories in a largely historical context. There are other various shifts that occur in parallel or intersect at a certain point, but the choice of these three prominent and predominant phases makes it easier to understand the practice of sound in Indian cinema in a coherent and comprehensive manner. This loose division also reflects the practical discourse of sound practitioners – in their vocabulary,[21] the various historical phases, as we see in their conversations, are named in similar nomenclatures such as 'direct sound era', 'mono era', 'dubbing era', and 'digital era'. Likewise, the tri-partite model works as a convenient lens to examine developments in the way sound is practiced in trajectories of Indian cinema, altering sonic and spatiotemporal experiences.

21 See interviews with sound practitioners in *Between the Headphones* (2021b).

Critical listening to film examples

After this historical outline, the later part of the book is devoted to a descriptive analysis of various modes and facets of sound production in Indian film and audiovisual media, ranging from mainstream popular Hindi language films made in Mumbai to regional cinemas in Bengali, Malayalam, and Tamil languages among others, as well as other various creative media productions. This comprehensive overview of sound production practices will be contextualized in the light of film and media production studies, the phenomenology of sound and listening, and sound studies via a critical listening to relevant film and audiovisual media examples as case studies.

A practice of attentive listening to film examples helps the examination of the trajectories of sound practice in Indian films and conceptualizes the use of sound since the advent of talkies to the contemporary digital realm. By studying and analyzing a number of film and media works produced in India from different technological phases of sound recording and design, the book establishes three corresponding models that are developed on the basis of historicizing sound practice and shifting sonic aesthetics in Indian cinema. These analytical accounts aim to critically engage with the conceptualization and realization of a creative practice with sound by way of squarely focusing on the film sound experience. The investigation makes a phenomenological survey of the sonic experience while making connections between various components of sound, from voice to sound effects and ambience. As methodology, this book doesn't focus on film music per se, though occasionally makes comments when songs or background music are intertwined with other creative layers of designed sound. More focused discussions on the uses of songs and music in Indian cinema can be found in other scholars' work (Booth and Shope, 2013; Morcom, 2016; Mukherjee, 2007; Arnold, 1991). This book addresses basic questions around the choice of sounds in significant sequences from Indian film and media works carrying narrative information and evocative nuances, re-ordering and positioning sound creatively to construct the mediated sonic environment. The book also tries to define aesthetic choices in the communication and cognition of sound in relation to film and media. The discussion and theorization by sound analysis deals with narrative discourses that arise from the different strategies of converging sound and moving image, as well as mixing particular sounds with other related sound components to make meaning. These analyses take their point of departure from specific phases of technological transitions and intend to highlight characteristics defining the sound aesthetics that emerge from these different phases of sound practice. The listening approach furthermore seeks to trace out the emergent aesthetics within theoretical frameworks of practice-led sound studies.

These analyses of film examples through different phases of cinematic sound practice aim to probe whether Indian films and media productions have shifted audio-audio and audiovisual relationships away from maintaining the merely vococentric contract of an early cinema soundtrack towards the spatial practice of a sonically present environment evident in digital production, in which the spectator's association with the film space is increasingly instigated by the spatial re-ordering of creative sound layering. This argument is substantiated by experiencing, examining, and dissecting a number of representative film and media works as case studies from three primary technological phases: analogue optical direct recording and mono-aural mixing; analogue magnetic recording, dubbing, and stereophonic mixing; digital multi-track sync recording and surround sound design. The selection of films discussed here as case studies cover a wide range of empirical materials for discussion, from the realist works of Indian auteur Satyajit Ray (1955–91) to the Indian indies as mentioned earlier (e.g., the contemporary Indian films of Anurag Kashyap and Dibakar Banerjee that are shot using digital sync recording imbibing a realistic approach to sound). Ray's entire oeuvre has been considered in terms of the agency of realism in the early monaural setting of Indian cinematic sound, revived in indie films. Further, a great number of relevant films from the analogue optical, analogue magnetic, and digital eras are referred to in order to locate specific characteristics of creative sound practice in the respective phases. Among audiovisual media works, projects that creatively incorporate sound are discussed in the light of current scholarships. Citing many examples from representative films and media works for critical listening, descriptive analysis, and scholarly reflection, the book provides an entry into the complex, fascinating, and ever-evolving sound world of Indian film and media, composing a valuable scholarly contribution.

Auto-ethnographic intervention

Correspondingly, this book learns about the nitty-gritty of sound production through in-depth conversations with prominent Indian sound practitioners as first-hand documentation of what has been achieved in sound use. This grounded knowledge works as a prerequisite in understanding how the mise-en-sonore (Chattopadhay, 2021a) or auditory setting is produced in Indian film and media. A discussion of the practice with a claim that sound production with digital technology has impacted cinematic experiences is thus based in solid empirical evidence. The conversations with the sound practitioners working in India are useful to enrich the bottom-up research of the book and ground it in the practical worlds of sound production in Indian film, audiovisual media, and the creative industries. These edited and collected interviews are published in a separate book *Between the Headphones: Listening to the Sound Practitioner* as a reference volume (Chattopadhyay, 2021b).

The conversations therein are based on a specific set of semi-structured and open-ended questions about the handling of sound from recording to design and how technology impacts these production processes. In the book, I discuss my dialogic approach to derive knowledge from practitioners about the intricate sound production process in detail. These are in-depth conversations rather than mere interviews from an outsider's position; an auto-ethnography approach is embraced, as I am a sound practitioner myself. These friendly conversations often depart from a set of semi-structured and open-ended questions to become candid dialogues navigating anecdotal evidence about production processes. According to scholars of qualitative research Jody Miller and Barry Glassner, semi-structured and open-ended interviews may solicit 'authentic accounts of subjective experience' (2011, p. 131). I have shown in *Between the Headphones* (2021b) that this approach helped the sound practitioners 'to speak in their own voices about their art and craft' (LoBrutto, 1994, p. 1). Sound Studies scholar Mark Grimshaw asserts that a questionnaire-based qualitative approach involving semi-structured interviews 'allows the interviewer a certain level of control which directs the interviewee down particular paths. Equally it allows the interviewee to expand on themes outside the limits of the questions, which can reveal unexpected information' (2011, p. 54). The dialectics between the top-down approach of reflective analyses and the bottom-up approach of learning from the practitioners themselves thus form the backbone of this book's research, ensuring that 'even the more abstract notions about filmmaking and cinema remain grounded in real-world practices' (Kerins, 2011, p. 10). Employing the approach of sonic auto-ethnography means that a self-reflective moment is helpful in unpacking the sound production stages and their intertwining histories and aesthesis. These illuminating inputs and evidential accounts, occasionally referred to from my other book *Between the Headphones* (2021), enriches this research monograph with practical insights based on the rudimentary aspects of the real world of sound production. This empirical evidence sheds light on India's normative modes of cinematic sound production and distribution chains. For example, according to one of the sound practitioners,[22] one reason for not using sync sound earlier had been the working structure within the Indian film industry, based on the stardom and inflexible power hierarchies and structural inequalities embedded in the systems in place. With the advent of digital technologies, sync sound started to require the overvalued actor's complete participation on the film set on an equal footing with the location sound technician, who has long held a lower status in film crew hierarchy. According to this empirical evidence, the introduction of digital technology has opened up scope for a more creative sound practice that not only has substantially changed the sonic experience but also realigned the hierarchy of the film crews, making the role of sound practitioners more significant and indispensable in India's new audiovisual environments.

Chapter Overview

The book is divided into 13 chapters, including this introduction. Most of the chapters contain sub-chapters with specific detail and in-depth discussion.

The introductory chapter divides sound production practices in Indian film and media into three primary phases. Critically listening to the trajectories of sound production and drawing ideas from current research in sound and music in Indian cinema (Booth, 2011; Morcom, 2016 et al), an observation is made that sound has been rendered in manifold methodologies and produced through various phases of production practices because of technological innovations and shifts. The dominant factors that have determined these shifts are rooted in technological developments (Kassabian, 2013; Kerins, 2011, 2006; Lastra, 2000; Altman, 1992). Certain aesthetic choices were made available by these technological phases – a process that shaped the historical evolution of sound practice in India. In this chapter, the historical trajectories are studied to understand the various differing and concurrent practices, methodologies, and approaches, as well as the shifts in recording and organizing sound for the purpose of audience engagement.

I divide these trajectories into three primary historical markers that seemed useful while locating and mapping the foremost technological shifts. I study how these shifts have become manifest in the emerging aesthetic choices available to and embraced by sound practitioners and how these choices are reflected in the production of a site's sonic presence. In other

22 See: interview with Anup Dev in *Between the Headphones* (Chattopadhyay, 2021b).

words, the various forms and formats of technological innovations and transformation have informed the usage of specific sound components. Therefore, my central concern is to study the specific nature of sound's usage in these corresponding and intercepting phases of recording and sound production in India: the analogue monaural synchronized sound recording (direct optical on location) and monaural mixing; dubbing, studio processed, and stereophonic mixing; digital multi-track synchronized recording and surround sound design. Critical observation, reflection, and analysis of the passages of sound from representative Indian film and audiovisual media works, specifically from these three different phases of sound production, are made. In light of the above, this book sets out to challenge the traditional reading of Indian cinema as producing mostly song and dance sequences. Questions might be raised as to whether this broad historical overview suffers from not being able to adequately delve into a focused and detailed discussion of a particular era. In its defence, I refer to the argument of certain film and media historians who maintain that writing history demands a broader perspective in order to accommodate historiographical accounts (Tybjerg, 2013). Emerging from a void of serious and sustained research on sound in the Indian films, this book could be viewed as creating a reference volume for future research that might present more detailed studies of particular periods of sound production. Moreover, the project is inspired by my own professional background as a sound practitioner. My academic experience in the historical developments of sound in Indian film and media arts help to open up the research to make new ranges of generative knowledge.

The second chapter historicizes the technological trajectories of film sound production from direct sound of early Indian cinema via the coming of magnetic recording and the primacy of dubbing to the advent of digital technology in film sound since 2000s, incorporating practices like sync sound recording and surround design.

The third chapter traces the various modes and facets of sound production in Indian cinema during the early talkies practicing direct sound. This chapter delineates the evolution of sound in Indian cinema from the silent era to sound film. Starting with the use of sound in early Indian talkies, such as *Bombay Talkies* (1934–53), the chapter elaborately describes the development of 'direct sound' and the resultant linguistic divisions in Indian cinema by analyzing a number of monaural works.

The fourth chapter studies sound in the so-called Golden Age of Indian cinema, when the quality and quantity of production sky-rocketed and reached a global stage.

The fifth chapter develops a careful study of the works of filmmaker Satyajit Ray. His work with sound is analysed to conceptualize what I term 'audiographic realism': a coinage that emphasizes the realistic sound elements recorded and incorporated in the monaural rendering of cinematic sound produced in this era. This chapter is an edited and reworked version of my article 'The World Within the Home: Tracing the Sound in Satyajit Ray's Films' published in *Music, Sound, and the Moving Image* (2018).

In the sixth chapter, I discuss the so-called 'dubbing era' from roughly the 1960s to the late 1990s. Many representative popular films from this era are examined to develop a basic argument that the creative practice of designing sound in Indian films of this period incorporated a technologically informed approach, using analogue sound processing with expressionistic and melodramatic overtones to cause the spectator to imagine film space instead of bodily experiencing it in film sound production. Parts of this chapter appeared previously in an edition of the *New Soundtrack Journal* entitled 'The Auditory Spectacle: Designing Sound for the 'Dubbing Era' of Indian Cinema' (2015).

In the seventh chapter, I critically discuss sound practices and sonic aesthetics through strands of the parallel film movement in India, focusing on the works of pioneering auteurs such as Mrinal Sen, Ritwik Ghatak, Mani Kaul, Shyam Benegal, Adoor Gopalakrishnan, G. Aravindan, Buddhadeb Dasgupta, Goutam Ghose, and Rituparno Ghosh, among others.

In the eighth chapter, following film analysis and finding actual evidence in conversation with sound practitioners, I demonstrate that stereophonic mixing, as an extension of magnetic recording and dubbing practices, rendered the cinematic imagination of this period as something spectacular, with extravagant songs and dances in exotic locations and action-filled scenes packed with studio-manipulated and synthetic sound effects. Add to this a deliberate lack of ambience, and this practice triggered a cinematic experience of emotive tension and affective stimulation.

In the ninth chapter, I study the advent of digital sync sound and the resultant (re-)emergence of spatial awareness in Indian cinema post-2000, termed the 'digital era'. Since the late 1990s, a large-scale conversion from analogue recording and analogue production practices to digital technologies has taken place in Indian cinema. Digital technology has been integrated into the production and post-production stages of filmmaking, as well as reproduction and projection formats. The ramifications of cinema adapting to a new technology have been far-reaching, particularly as a result of novel digital sound practices in Indian cinema. Production practices and techniques, such as location-based, multi-track synchronized recording and surround sound spatialization, altered the notion of the film soundtrack in the contemporary digital realm of Indian cinema. This subchapter carefully studies how the processes of digitalization made a substantial impact on the narrative strategies and aesthetic choices of extant cinematic sound production, informing the creation of the presence of a site in the film space by novel modes of interplay between quasi-diegesis and mimesis.

The tenth chapter expands the discussion of the digital realm of sound production in Indian cinema towards the surround revolution. It discusses representative films using spatial audio from 5.1 to the present era of Dolby Atmos, studying the site-specificity and spatial awareness such films encourage. An earlier version of this chapter appeared in an edition of *Journal of Sonic Studies* entitled 'Being There: Evocation of the Site in Contemporary Indian Cinema' (2016).

The eleventh chapter is dedicated to the study of the creative practice of sound in other audiovisual media art productions such as interactive installations, works for digital media arts, and sound arts. Several artworks, projects, and initiatives are discussed.

The twelfth chapter starts with a discussion of feminist filmmakers in India and their contributions. The chapter then elaborates on the independent films of the digital era, termed the 'Indian indies', following the so-called 'parallel cinema' and its inclination toward audiographic realism expanded with the use of sync sound and location recordings. Later, the chapter makes a critical overview of the use of sound in documentary, shorts, experimental, and art-house cinema genres of India and regional films. The aim is to examine the processes of sound recording and design, and the modulation of the reality, widening the scope of experimentation in sound and audiovisual media.

Based on an in-depth and historical enquiry made into the various threads of sound production in India – by means of critical listening, descriptive analysis, and informed commentaries – in the thirteenth and final chapter, I consolidate the content and the context of the book and its focus on the uniqueness of sound practice in India, its historical trajectories, as well as aesthetic shifts by formulating concluding remarks. As already discussed throughout the book, Indian cinema – with its diverse fields of practices, productions, and experiences – is arguably the world's largest producer of films. It is also an emerging giant in producing a diverse range of other audiovisual media content. It is the intention of this chapter to locate the dominant tendencies and predilections of this diversity with a practice-aware approach. This approach is driven by the methodologies of critical and self-reflective observation, pursued through a number of the complex and intercepting threads of production trajectories in a vastly heterogeneous national cinema, loosely unified by technological and aesthetic developments. Being a sound practitioner myself, my critically observational and reflective approach helps to creatively locate the major historical developments in the practice of sound, which is in itself an underexplored subject in film history and media studies, and screen music research, as well as within the emerging field of sound studies. My aim is also to facilitate a study of film sound that questions the classical assumptions of the sound-image relationship and shift the focus towards the presence of the diegetic space and mediated universe as a vital narrative and creative component. I have devised a taxonomical model based on critical observations of historical and technological shifts and emergent aesthetic strategies. This model's point of departure is found in specific phases of technological innovations and transitions in sound production but is not limited to a discussion of the history of sound technology. On the contrary, the model highlights characteristics delineating sonic aesthetics emerging from these prominent technological phases, thereby linking various modes of sound practice with the varied fields of film and media studies, as well as screen music production.

References

Altman, R. (ed.) (1992). *Sound Theory/Sound Practice*. New York: Routledge.

Arnold, A. (1991). *Hindi filmī gīt: On the History of Commercial Indian Popular Music* (Ph.D. Dissertation). Urbana-Champaign: University of Illinois.

Biancorosso, G. (2009). "Sound". In Plantinga, C.R and Livingston, P. (eds), *The Routledge Companion to Philosophy and Film* (pp. 260–267). London: Routledge.

Blesser, B. and Salter, L.R. (2009) *Spaces Speak, are you Listening?: Experiencing Aural Architecture*. London: The MIT press.

Bloom, Peter J. (2014). "Sound Theory," in Edward B. and Warren B. (eds), *The Routledge Encyclopedia of Film Theory* (pp. 430–434). London and New York: Routledge.

Booth, Gregory D. and Shope, B. (eds) (2013). *More Than Bollywood: Studies in Indian Popular Music*. New York: Oxford University Press.

Born, G. (ed.) (2013). *Music, Sound and Space: Transformations of Public and Private Experience*. Cambridge, U.K.: Cambridge University Press.

Branigan, E. (1989). "Sound and Epistemology in Film". *The Journal of Aesthetics and Art Criticism* 47(4), 311–324.

Bull, M. (2018). *The Routledge Companion to Sound Studies*. Abingdon, Oxon; New York, NY: Routledge.

Chattopadhyay, B. (2007). "Sound Memories: In Search of Lost Sounds in Indian Cinema". *Journal of the Moving Image* 6, 102–111. Jadavpur: Jadavpur University press.

Chattopadhyay, B. (2014a). "Object-Disoriented Sound: Listening in the Post-Digital Condition". *A Peer-reviewed Journal About* 3(1). www.aprja.net/?p=1839 (accessed 13 April 2023).

Chattopadhyay, B. (2016). "Being There: Evocation of the Site in Contemporary Indian Cinema". *Journal of Sonic Studies* 12. https://www.researchcatalogue.net/view/286592/286593 (accessed 12 September 2023).

Chattopadhyay, B. (2017). "Reconstructing Atmospheres: Ambient Sound in Film and Media Production". *Communication and the Public* 2(4) 352–364. London: SAGE Publication.

Chattopadhyay, B. (2018). "Orphan Sounds: Locating Historical Recordings in Contemporary Media". *Organised Sound* 23(2), 181–188. Cambridge: Cambridge University Press.

Chattopadhyay, B. (2021a). *The Auditory Setting: Environmental Sounds in Film and Media Arts*. Edinburgh: Edinburgh University Press.

Chattopadhyay, B. (2021b). *Between the Headphones: Listening to the Practitioner* (2021b). Newcastle upon Tyne: Cambridge Scholars Publishing.

Chion, M. (1994). *Audio-vision: Sound on Screen*. Translated and Edited by Gorbman, C. New York: Columbia University Press.

Dyson, F. (2009). *Sounding New Media: Immersion and Embodiment in the Arts and Culture*. California: University of California Press.

Gopalan, L. (2002). *Cinema of Interruptions: Action Genres in Contemporary Indian Cinema*. London: British Film Institute.

Ganti, T. (2012). *Producing Bollywood: Inside the Contemporary Hindi Film Industry*. Durham, North Carolina, United States: Duke University Press.

Hayward, S. (2006). *Cinema Studies: The Key Concepts*. London: Routledge.

Ihde, D. (2007). *Listening and Voice: Phenomenologies of Sound*. New York: Sunny Press.

Jordan, R. (2012). "The Ecology of Listening While Looking in the Cinema: Reflective Audioviewing in Gus Van Sant's Elephant". *Organised Sound* 17(3), 248–256.

Kassabian, A. (2013). "The End of Diegesis as We Know It?" In John Richardson, Carol Vernallis, and Claudia Gorbman (eds), *The Oxford Handbook of New Audiovisual Aesthetics*, pp. 89–106. Oxford: Oxford University Press.

Kerins, M. (2011). *Beyond Dolby (Stereo): Cinema in the Digital Sound Age*. Bloomington: Indiana University Press.

Kerins, M. (2006). "Narration in the Cinema of Digital Sound". *The Velvet Light Trap* 58 (Fall), 41–54.

Kaul, M. (1977). "Communication". Symposium on THE CINEMA SITUATION, with Kumar Shahani, Adoor Gopalakrishnan, Shyam Benegal and Dileep Padgaonkar. https://theseventhart.files.wordpress.com/2011/07/the-cinema-situation_74.pdf (accessed 4 July 2022).

Kaul, M. (1983). "Towards a Cinematic Object". *Indian Cinema Super Bazaar.* https://theseventhart.files.wordpress.com/2011/07/towards-a-cinematic-object_mk.pdf (accessed 4 July 2022).

Kaul, M. (1991). "Seen from Nowhere". In Vatsavan, Kapita (ed.), *Concepts of Space: Ancient and Modern*. New Delhi: Abhinav Publications.

LaBelle, B. & Martinho, C. (eds) (2011). *Site of Sound: Of Architecture and the Ear: Vol 2*. Berlin: Errant Bodies Press.

Lastra, J. (2000). *Sound Technology and the American Cinema: Perception, Representation, Modernity.* New York: Columbia University Press.

Morcom, A. (2016). *Hindi Film Songs and the Cinema*. Farnham, UK: Ashgate Publishing.

Mukherjee, M. (2007). "Early Indian Talkies: Voice, Performance and Aura". *Journal of the Moving Image* 6, 39–61. Kolkata: Jadavpur University Press.

Nancy, J. L. (2007). *Listening.* Translated by Charlotte Mandell. New York: Fordham University Press.

Pinch, T. and Bijsterveld, K. (2013). *The Oxford Handbook of Sound Studies*. Oxford: Oxford University Press.

Rajadhyaksha, A. (2007). "An Aesthetic for Film Sound in India?" *Journal of the Moving Image* 6. https://www.jmionline.org/article/an_aesthetic_for_film_sound_in_india (accessed 7 September 2023).

Ray, S. (1976). *Our Films, Their Films*. Delhi: Orient Blackswan.

Ray, S. (2011). *Satyajit Ray on Cinema*. Edited by Sandip Ray. New York: Columbia University Press.

Rogers, A. (2013). *Cinematic Appeals: The Experience of New Movie Technologies*. New York: Columbia University Press.

Sterne, J. (2012). "Sonic Imaginations". In Sterne, J. (ed.), *The Sound Studies Reader* (pp. 1–17). London: Routledge.

Sterne, J. (2006). "The Death and Life of Digital Audio". *Interdisciplinary Science Reviews* 31(4), 338–348

Sterne, J. (2003). *The Audible Past: Cultural Origins of Sound Reproduction*. Durham: Duke University Press.

Théberge, P., Devine, K., and Everrett, T. (2015). *Living Stereo: Histories and Cultures of Multichannel Sound*. New York: Bloomsbury.

2. The Technological Frameworks

Abstract: The second chapter historicizes the technological trajectories of film sound production from early Indian cinema's direct sound to the advent of digital technology, locating and critically charting the film industry's foremost technological innovations and shifts: monaural synchronized sound recording; dubbing, and stereophonic mixing; digital multi-track synchronized recording and surround sound design. The chapter studies how these shifts manifested in emerging aesthetic choices embraced by sound practitioners and how these approaches are reflected in the production of sonic environments and presence. In other words, the various forms and formats of technological innovations and transformation have informed the usage of specific sound components. Central is the specific nature of sound's usage in these corresponding and intercepting phases of recording and sound production in India.

Keywords: monaural recording, direct sound, stereophony, digital technologies, sync sound

Optical Recording and Direct Sound

The earliest sound recordings in India were registered in 1902, when London's Gramophone and Typewriter Ltd. sent its recording engineers for an ethnographic expedition in South Asia to gather music and sound recordings. The intention of the company was to set up a sound industry in the colony in order to export the profit back to Britain through the exploitation of local resources, labour, and materials. Indian films were still silent when these developments took place. The gramophone had a separate trajectory of technological transmission in India, as did radio. Whereas the era of 'talkies' began in the United States with the release of *The Jazz Singer* in 1927, sound came to the still nascent Indian film industry only after investors and studio owners travelled to the United States to experience this new phenomenon. Once these producers, such as

Chattopadhyay, B., *Sound in Indian Film and Audiovisual Media: History, Practices, and Production*. Taylor & Francis Group, 2024
DOI 10.5117/9789463724739_CH02

Jamshedji Framji Madan of the Madan Theatre, experienced sound film, they committed to bringing this technology to India. By 1931, the first Indian sound film, *Alam Ara* (*Ornament of the World*), directed by Ardeshir Irani, was released along with a large number of other films in Bombay, Kolkata, and elsewhere – all talkies.

The gramophone, and later film and radio, all modern inventions imported from the West, irrevocably altered sound practices in India in the early twentieth century. With the coming of sound recording technology, major unsettling transitions took place, and filmmaking moved from the hand of amateurs to the bigger studios that could afford the technologies of sound recording and reproduction. These early sound recordings in Indian films were all on direct optical recording using sound-on-film technology, meaning that image and sound were recorded with bulky equipment directly onto the optical filmstrip simultaneously with the actor's performance without any post-synchronization. Indian film scholars Madhuja Mukherjee (2012) and Olympia Bhatt (2019) show how the coming of sound forced the film industry to embrace certain changes: many actors lost their jobs because their voices weren't considered appropriate as language-based regional nationalism emerged. These socio-economic changes are the subject of film-sociological studies. From an interest in sound practice and aesthetics, I will focus squarely on sound production proper and briefly touch upon technological shifts.

While *Alam Ara* (Ardeshir Irani, 1931) is widely known as the first sound film made in India, it is a lesser-known fact that, in the same year, roughly 26 more Indian sound films were produced in different languages, including Tamil, Hindi, and Bengali (Rajadhyaksha, 2016). These early sound films involved a few prominent American and European practitioners such as Franz Osten, Josef Wirsching, and Wilford Deming Jr.

Deming Jr., contributed substantially to the technological development of early Indian talkies as a transfer of knowledge from Hollywood to India's emerging film industry. Deming Jr. left a series of essays delineating the deplorable conditions in which early Indian sound films were made. In his essay 'Talking Pictures in India' published in *100 Years of Cinema*, Deming notes on a film shooting:

> [A] bit of trouble was experienced with noisy microphones, ...several retakes were necessary. Dynamic microphones seem the only answer to all year operation in India. One interesting phase of Indian work is the complete indifference with which the microphone is received. (1995, p. 366)

Deming's serious charges about the quality of sound in Indian film productions weren't unfounded. *Alam Ara* and the other early Indian talkies used direct optical recording and monaural reproduction for the first time. In most of these productions, sound was not given much attention. Due to a lack of funds, director Ardeshir Irani couldn't afford hiring Deming's service in *Alam Ara* and so did the sound himself (ArZan, 2019). One of the earliest optical recording machines used in Indian cinema was Fidelitone, recording on two channels in the sound-on-film method. Around 1933, Visatone was incorporated. This was another sound-on-film recording machine, which ran on 220 volts and in 50 cycles per second, with two-channel recording. For Indian sound technicians, this was a time for learning the craft of sound, experimenting with available foreign sound equipment imported from the colonial West, and gradually standardizing the sound production process through trial and error. As part of this experimentation, in a particularly decolonial spirit, a few local innovators made their own recording machines with limited resources. In 1935 Bamadas Chattopadhyay built a homemade recording machine called the Csystophone that was used in a few Bengali-language films.[1] Conversations with veteran sound practitioners (Chattopadhyay, 2021b) reveal that due to a lack of resources and adequate funding such home-grown endeavours couldn't be sustained (Bhattacharya, 2009).

The following period, until at least the late 1950s, was an era in which Indian cinema adapted to and appropriated the technicalities of direct optical recording using novel sound gear such as the microphone and monaural reproduction. It was no surprise that these technological innovations and the increasing skill to master them coincided with the increasing political mood to push for independence from British rule. The directly recorded sound in these films tended to provide for a mediated but realistic representation of the pro-filmic spaces as India struggled for independence. With the advent of new cameras and sound recording equipment, keeping traces of a series of mass protests from the anti-colonial *Swadeshi* (of one's own land) movement, which encouraged the use of domestic Indian produce, aided by a resurgent political life in India, was possible.[2] The erstwhile and almost forgotten filmmakers Hiralal Sen and Motilal Sen documented Sir

1 See: Anonymous, "Csystophone". In *Sound Cultures in Indian Cinema*, Item #11, http://sounds.medialabju.org/items/show/11 (accessed May 4, 2021).

2 *Swadeshi* was a significant movement in Bengal, which advocated for the use of domestic (*swadeshi*) products including cotton and other cloth made in India. The *Swadeshi* movement was a moment of resurgent Indian nationalism against British rule.

Surendranath Banerjee's political lectures in Kolkata.[3] Later, Haren Bose (father of Nitin Bose) recorded Banerjee's lectures that were banned by the British authorities.[4] However, with the coming of sound in the hands of film businessmen, many of whom were potential British allies, entertaining music and songs became the most identifiable sounds for talking films produced in different parts of India. This non-diegetic music and these narrative-interrupting songs were recorded live in pro-filmic spaces with musical instruments and musical performers who were hidden behind the camera in mobile vans, as playback systems hadn't yet been invented. This method posed several questions such as that of a make-belief diegesis and audio leakage in the form of incidental and ambient sounds entering the film space unknowingly through the open microphone on location – I will discuss these issues in a later chapter. In line with the promise of realism, combined with opening up the possibility of qualitative digressions in the hands of profit hungry film studios and businessmen, the monaural practice of sound production and reproduction continued during the 1950s and 1960s, a period considered to be the Golden Age of Indian cinema, when film auteurs such as Satyajit Ray and Chetan Anand emerged and placed Indian cinema on the world stage, and film directors like Guru Dutt, K. A. Abbas, and Bimal Roy helped to set the benchmarks in the craftspersonship of Indian filmmaking along with an acquired and seasoned sensitivity towards contemporaneous issues and socially vibrant sites. These developments in Golden Age sound practice will be discussed in a following chapter.

Direct sound generally refers to the technique of recording the music, voice (dialogue and otherwise), ambient, and incidental sounds of a scene at the moment that the pro-filmic space is being captured on film. Since the introduction of synchronized sound in the film *Alam Ara* (Irani, 1931), sound practice in India was monaural from recording to mixing stages and to reproduction, meaning that, 'a single channel of sound was played from a loudspeaker placed behind the screen, creating the illusion that the sound of the film was emanating from the projected images' (Kuhn and Westwell, 2014). I have thoroughly discussed the aesthetics of early monaural production and the reproduction of film sound and the resultant tension in narrative mounting in *The Auditory Setting* (2021a), where my

3 Surendranath Banerjea (1848–1925) was one of the earliest Indian national leaders fighting against British imperialism. He was a skilled orator and travelled all over India to give speeches. He later became a member of the Indian National Congress.

4 As learnt from an interview with Santu Basu, yesteryear's sound recordist from the publication *Dhanimoyotar Itibritto* (Roookala Kendro, Kolkata, 2009).

concern was to study the use of ambient sound in global cinema following different film historical trajectories. In that book, I note that as early as 1928, Rudolf Arnheim recognized the problems of the single-source and screen-centric practice of monaural synchronized sound, as it was doing away with the significant interplay between the division of the picture and three-dimensional movement within space (see Bloom, 2014). Arnheim meant that 'synchronized sound distracts from the significant play of visual interpretation among all the elements of the image and instead locks the viewer into the space' (quoted in Bloom, 2014, p. 431). I argue that such criticism arose due to the overarching emphasis on the 'marriage'[5] of moving image and sound source in one speaker, as the monaural source placed behind the screen was engineered according to standards of monaural sound production and reproduction. I show how today's surround-sound environment allows the source more freedom of movement outside the screen-centric coupling between sound and image, as Arnheim wished. The off-screen three-dimensional sensibility he looked for in cinema is realized today in the spatial environment of surround-sound but was limited in scope with synchronized monaural (direct) recording and monophonic mixing due to their screen-centric renderings of sound and image. Arnheim's problem with monaural aesthetics of synchronization emphasized 'the differences between film and reality as a key artistic quality of film form' (quoted in Bloom, 2014, p. 431). As such, Arnheim was concerned that the synchronization of monaural practice reduced three-dimensional experience onto the two-dimensional screen, disturbing film's relationship to reality.

Film theorist Béla Balázs's ideas on sound film contrast strongly with his contemporary Arnheim. As I noted in *The Auditory Setting* (2021a), his defence of synchronized sound and monaural aesthetics was based on an extension of cinema's narrative capacities, since the experience of sound becomes more spatially defined. Peter J. Bloom refers to Balázs's stance: 'The potentially spatial characteristics of sound, which Altman (1992) has further described as the "material heterogeneity" of sound, may then be better guided, Balázs insists, through a visual representation. The image assists on disentangling the location of voices speaking, for example, as attached to different speakers appearing on screen with their own distinct qualities and physiognomies of expression' (2014, p. 433).

Film sound scholar James Lastra interprets these two different viewpoints as represented by Arnheim and Balázs describing two corresponding 'models'

5 'Married print' has been a standard term in the film industry to denote the combining of sound and image on a single irreversible optical print.

operating within classical film narrative in Hollywood and the construction of its sonic spaces:

> The first, heir to metaphors of human simulation and described in terms of perceptual fidelity, emphasizes the literal duplication of a real and embodied (but invisible) auditor's experience of an acoustic event. Its watch-words are presence and immediacy. ...Aesthetic perfection entails the absolute re-presentation of the original in its fullness so that no differences whatsoever exist between the pro-phonographic experience and the phonographic one (2000, p. 181)

The other model, according to Lastra, 'emphasizes the mediacy, constructedness, and derived character of representation' (2000, p. 181). These two opposing stances frame a similar tension operating in Indian cinema. On the one hand, narrative storytelling pleasure is provided by a primarily musical and vococentric representation of sound (Chion, 1994), exemplified by the many religious and devotional films of early 1940s and late 1950s Indian films, and later championed by the song and dance routines of syncretic *Bollywood*. On the other hand, the realistic re-presentation of actual sites, actors, and social situations is rendered through direct recording and synchronized sound practices, as demonstrated by social realist films of the 1950s such as *Neecha Nagar* ('Lowly City', Anand, 1946) and *Do Bigha Zamin* ('Two-Thirds of an Acre of Land', Roy, 1953). These two polarities defined Indian cinema when auteurs like Satyajit Ray emerged. Ray, by his authorial choice, embraced and advocated for the latter model and an alternative style of Indian cinema based on social realism, which was followed by the Parallel Film movement[6] in the 1980s and 1990s and later with the 'Indian indies' (the Indian Independent Cinema movement) in the 2000s.

Such a reading of Indian cinema is based on various methodologies and conceptual approaches to sound practice and sonic aesthetics. This

6 Parallel Cinema was a film movement in Indian cinema, an alternative to the mainstream commercial Indian cinema, especially represented by the popular Hindi cinema genre known as Bollywood. Inspired by Italian Neorealism (particularly works by Vittorio De Sica), Parallel Cinema began just before the French New Wave and was a precursor to the Indian New Wave of the 1960s and Indian indies in the 2000s. The Parallel movement was initially led by Bengali cinema, spearheaded by internationally acclaimed filmmakers such as Satyajit Ray, Mrinal Sen, Ritwik Ghatak, Tapan Sinha and others. It later gained prominence in other film industries of India and Bangladesh. Parallel Cinema is known for its serious content, realism and naturalism with keen ears for the sociopolitical conditions of the times, and as a challenge to the inserted song-and-dance routines that are typical of mainstream Indian films.

reading helps to understand the use of the auditory details that social realist films offered in terms of specific sound practice. In this practice, there was a primacy of locative recording and synchronized sound as opposed to obviously post-synchronized sonic aesthetics. This latter aesthetic strategy manifested in the popular song and dance routines or in the asynchronously made mythological and devotional films that dealt with subjects beyond ordinary reality. The result of the former locative social realist approach was a sonic experience that relied on perceptual fidelity and aesthetic perfection. This required a faithful re-presentation of the pro-filmic spaces in terms of a screen-centric use of location-specific sounds and actors' live voices with monaural technique of sound recording and reproduction.

With this context in mind, in a later chapter I discuss how the treatment of sound in Ray's and many early Parallel Cinema films made with a direct sound technique highlighted a distinct recognition of the presence of the site in the diegesis. This process involved listening to a locative social situation as a context for the narrative development in the synchronized 'direct' sound recording and monaural organization of sound. This practice created the precedence for a realistic auditory setting later championed by Indian Parallel Cinema filmmakers like Shyam Benegal, Govind Nihalani, and Adoor Gopalakrishnan in the 1970s and 1980s. These filmmakers' methods and narrative strategies of direct location recording, and detailed use of site-specific sounds are recognized, embraced, and revived in the 'digital era' from 2000s onwards in the Indian independent film movement. The emergence of multi-track 'sync' sound (Chattopadhyay, 2016) is incorporated into the Indian independent films of Nagesh Kukunoor, Dibakar Banerjee, Anurag Kashyap, Zoya Akhtar, Neeraj Ghaywan, Vikramaditya Motwane, and Hansal Mehta with a gritty social realist approach.

Central in this practice of direct sound is the methodology of telling a story in Indian cinema through the technology of (mostly direct) sound recording from the pro-filmic space in the relatively shorter period of optical recording direct-on-film and monaural reproduction (1930s–1950s). Following a number of films, in the third chapter of this book I investigate the role of actual site-specific and incidental sounds gathered during direct recordings to create a sense of the site's presence as aural situated-ness in the portrayal and construction of pro-filmic spaces. I show that this practice created a screen-centric mise-en-sonore or auditory setting in order 'to place the auditor as literally as possible in the pro-filmic space' (Lastra, 2000, p. 182). Using synchronized sound techniques and employing monophonic mixing inclined toward the direct sound methodology, this era of Indian filmmakers allowed the audience to experience the cinematic sites and characters inhabiting this pro-filmic space with vivid sonic information. This strategy added to the sense of believability in the story-world as expected from a familiarity with the places shown and delineated in the narrative, so that the audience was informed about the embodied presence of the site in the narration of the diegetic story-world (Burch, 1985; Birtwistle, 2010) by 'letting the camera be the eye, and the microphone the ear of an imaginary person viewing the scene' (Maxfield quoted in Lastra, 2000, p. 183). Following this approach driven by direct sound, early Indian films, like their Hollywood counterparts, 'simulate the perceptions of an observer located on the film set, whose eyes and ears (camera and microphone) are joined as inseparably as those of a real head' (Maxfield quoted in Lastra, 2000, p. 183), as in the Bombay Talkies productions that I discuss in the next chapter. These are the essential tenets of synchronized recording and monophonic production aesthetics, extended into Indian cinema with its own customs of traditional storytelling and a sociality imbued with vibrant song-and-dance cultures, community spirit, and the urge for social cohesion.

Although film music is not an explicit focus of this book, from a sound studies perspective, a critical attitude may be useful regarding the pressence of background music in commercially oriented mainstream Indian films. Such uses are aimed at grossly underlining moods and emotions as an affordable technique for audience engagement. Film music, in most film-making cultures, is used 'largely to set mood or elicit a particular emotional response from the audience' (Kuhn and Westwell, 2014). Film music tracks and sound effect tracks 'establish a particular mood' (Doane, 1985, p. 55) instead of providing a sense of space. In commercial and populist Indian films produced with the playback method (i.e., recording songs in the studio and playing them back on the film set for actors to lip sync to), this lack of spatiality through the overabundant use of non-diegetic music is observed. While direct era films were all synchronized with the music added on location, the playback method, starting with *Bhagya Chakra* (Nitin Bose, 1935), altered this mode of sonic site-specificity with the excuse of creating clarity, control, and entertainment.

With this trend in place, the hierarchy of different sound components started to follow certain conventions in the mixing stages of Indian films somewhat similar to Hollywood musicals: 'Sound effects and music are subservient to dialogue and it is, above all, the intelligibility of the dialogue which is at stake, together with its nuances of tone' (Doane, 1985, p. 55). In this hierarchy, site-specific, spatial, and incidental sounds remained fluid and flexible for creative intervention. It is, however, the very component of site-specific sound that can provide information about a site and the spatial context in the construction of the pro-filmic space – other components like effects and music are used as mere tropes for storytelling. As some sound practitioners suggest (Chattopadhyay, 2021b), layers of site-specific sounds, known as 'ambience', inject life and substance not only into what we see on screen but also into the off-screen diegetic world of Indian films. In the direct sound era (i.e., the optical sound-on-film method), sounds such as forest murmurs, rain, traffic roar, public gatherings, could enter the screen space and were able to provide spatial evidence to create a sense of aural situated-ness. The unintentional use of such sounds in Indian cinema's direct sound era could create a convincing, and realistic diegetic world. I show in a later chapter how the advent of the playback system destroyed this sense of aural situated-ness, bringing emotional manipulation to the foreground as a narrative strategy, and how this trend continues in the present.

Magnetic Recording and Dubbing

During the late 1960s, magnetic recording and mixing started to be extensively used in Indian cinema. The technology of recording on tape was invented in Germany[7] and used selectively in German national radio broadcasts. But after World War II, tape became widely incorporated into music and cinema production worldwide, including in India, gradually replacing the once-ubiquitous optical recording medium. Magnetic tape provided major improvements in audio quality, with a better signal-to-noise ratio,[8] better dynamic range,[9] wider headroom,[10] and a far better frequency response. Medium-specific noises like 'hiss' and 'hum' were

7 Magnetic tape recording was developed in Germany during the 1930s at BASF chemical company in cooperation with the Reich Broadcasting Corporation.

8 Signal-to-noise ratio (SNR) is a measurement used to describe how much desired sound is present in a sound recording, as opposed to unwanted sound (noise). This nonessential and undesired input could vary from static electronic noises like hum to the internal noise of the recording equipment, or incidental ambient sounds, such as the rumble of outdoor traffic, or the voices of a crowd in the background.

9 Dynamic range is the ratio of the loudest undistorted sound to the quietest discernible sound, expressed in decibels (dB), that an audio system is capable of producing. The compact disc's (CD) dynamic range is about 90 decibels compared to better-quality phonograph discs (LP), whose dynamic range is about 70 decibels.

10 Headroom in sound technology refers to the amount by which the signal-handling capabilities of an audio system exceed a designated level and is understood as a safe zone allowing transient audio peaks to exceed the nominal level without distortion.

introduced in place of the 'clicks' and 'pops' of the shellac disc and the sticky running sound of optical film sprockets. From 1950 onwards, magnetic tape quickly became the standard medium of sound recording in film industries across the globe, as well as in the radio and music industries, and gradually led, in the 1960s and 1970s, to the development of high fidelity (hi-fi) stereo sound and multi-track tape recording for music, replacing vinyl shellac discs as the primary mastering medium for sound production and distribution.

Magnetic tape as a recording medium contributed to a re-organization of the recording process. It extended recording times beyond the three-minute recording time constraint of shellacs and cylinders, and offered sound practitioners, as we learn from interviews (Chattopadhyay, 2021b), a sense of plasticity and malleability. Sound recorded on tape could then be edited, joined, and manipulated sonically in ways that weren't possible in optical recordings made directly on film. With the advent of the magnetic medium, it became possible to clean, erase, overdub, and employ multi-track mixing. Following the emergence of magnetic recording, in-house sound studios became popular for doing post-production instead of direct recording on location. As a result, film sound became increasingly distanced from its real site narrated on screen. Gradually, an analogue studio-centric technique of film sound design emerged as the dominant mode of practice. Dubbing and Foley followed technological advancements with the introduction of the Nagra portable recorder,[11] a battery-operated or electrically powered, portable, professional audio tape recorder produced by Kudelski SA, based in Switzerland. utilizing quarter-inch magnetic tape, and the MagnaTech Rock-and-Roll mixing console, which is a mixing console revolutionizing tape-based innovations in film sound production, for example, looping, dubbing, ADR (Automated Dialogue Replacement) and re-recording.[12] With the advent of these technologies and machines, the studio system invited more investment in sound post-production. Tools and techniques like multi-track mixing, tape-based editing, loop dubbing, track-laying, and re-recording opened up the possibilities of the parallel resourcing of sound reconstruction, making a dependence on direct location recording obsolete. Stock sounds were shared or became commercially available as a bank

11 The Nagra II model has been the most popular of the series for use in the Indian film industry. I started my filmmaking career recording sounds with the Nagra II.

12 Sound designer Dipankar Chaki speaks of the early years of the MagnaTech Rock-and-Roll mixing console in the Bengali film industry and how that improved the sound quality. See: interview in *Between the Headphones: Listening to the Sound Practitioner* (Chattopadhyay, 2021b).

of 'sound objects' from which a sound editor could pick up raw materials for ambience and sync sound effects – although, in most cases, ambience was a minor concern in sound organization. With the rise of analogue magnetic recording, the sound objects recorded on magnetic tape were no longer referred to as a site-specific sound source (Demers, 2009). The practice of recording sound on tape enhanced sound's materiality, making it possible to decouple sound from the site as a raw material. At this point, one could bring these raw materials into the studio to follow processes of analogue post-synchronization and re-couple it with the image. Following this method, films were increasingly shot on a pre-designed set inside studios instead of at real locations, and film sound tended to become a mere dialogue-background, score-sync effect scheme. Thus, a practice was carried out by industry-dependent and technologically informed sound technicians to design a soundtrack for a film out of asynchronous sound sources, using pre-recorded sound materials. In most cases, those technicians did not give much attention to the spatial aspects of sound, which would otherwise demand closer attention to the location or the source of sound in its recording and design stages. Song-and-dance sequences with loud background music were used merely to mask these shortcomings[13] Many sound practitioners admitted that the use of a loud mix in the background score limited the possibility of carefully and creatively designing the soundtrack. The lack of creativity in sound recording and design meant a lack of information about the pro-filmic space in the narration. This trend of a sterile, studio-centric sound production gradually approached relative abstraction of the Indian film's diegetic space in the ensuing dubbing era.

After playback and dubbing were introduced, sound in Indian cinema was rarely recorded on location even less often. Dubbing and re-recording gained momentum as the specific mode of sound practice in popular films from this time, while the film soundtrack was mostly created (or re-created) in the studio. Actors would recite and re-record their lines as their images appeared on the studio screen in a process known as 'looping'[14] or ADR (Automated Dialogue Replacement). Background music, along with lavish song-and-dance sequences followed this, and various sound effects, known

13 See: interview with Dev in *Between the Headphones: Listening to the Sound Practitioner* (Chattopadhyay, 2021b).

14 See: interview with Anup Mukherjee in *Between the Headphones: Listening to the Sound Practitioner* (Chattopadhyay, 2021b).

as Foley,[15] entirely made in post-synchronization within the close confines of a studio, were added later as the secondary or tertiary layers of sound organization. I have shown in *The Auditory Setting* (2021a) how dubbing or ADR exacerbated this sense of spatial a-synchronism, along with an estrangement between voice and body.

Another technological innovation that pushed sound production further away from the location and into the studio was the popular Arriflex handy camera. The use of ADR in Indian cinema became a regular practice from the 1960s on with the arrival of Arriflex IIC and Arriflex III cameras, which required a blimp (a soundproof cover) to shield their notorious motor noise during location shooting.[16] This distracting camera noise required that all production sound had to be re-created in the studio. Eventually, this process became the standard practice in Indian films. As dubbing emerged alongside the standardization of analogue magnetic recording and mixing, it was facilitated by multi-track re-recording in the studio. The following phase of sound production in Indian cinema was shaped toward what is known as the 'dubbing era' (roughly between the 1960s and 1990s). This was a long stretch of time that illustrated a growing interest in the controlled deployment of a few sound elements as design materials in films, keeping the primacy of the voice. However, in this hierarchy of sound organization, there was a substantial lack of site-specific sounds and creative sonic practice. This approach was a result of the standardizing methods of studio-centric film sound production with the particular aesthetics of post-synchronization using dubbing and Foley, as explained above. Why was it that there was a lack of nuanced and artful sound practice in this era of Indian film production? What was the nature of diegesis as the narrative process used in this specific practice of spatial sound (or a lack thereof)? And how was the presence of cinematic site produced in this phase of Indian cinema? To address these questions, I will examine the specific sound practice of 'dubbing' as a cinematic technique embraced in this period by sound practitioners. Magnetic recording, dubbing, re-mixing, and re-recording instigated a technologically simulated approach to represent reality in overly expressionistic, loud, and

15 'Foley' is a technique used in the sound post-production stages of filmmaking to recreate sound effects of a scene for post-synchronization with the image. It allows for clean recordings of the effects synthetically made inside a studio, since production mics would include unwanted locational noises, which are, of course, almost entirely absent in the studio. Foley is traditionally used to provide the actor's footsteps, movements, and other 'personal' sounds, all made post-synchronously.

16 For elaborate analysis of the term, see: *The Auditory Setting: Environmental Sounds in Film and Media Arts* (Chattopadhyay, 2021a).

melodramatic overtones that distanced sound's actuality away from the location, using loud mixes of voice and music aimed at creating emotional responses from audiences, gradually giving rise to the studio's control over sound production. In earlier writing (Chattopadhyay, 2012c, 2013, 2015), I underline this tendency.

Magnetic recording and its development in dubbing and re-mixing in popular Indian cinema rendered this a-synchronous practice into what Sergi, in a Hollywood context, describes as 'spectacular' (Sergi, 2004; Kerins, 2011). In Indian cinema this spectacle was rather sonically manipulated: an expanded, fantastical experience with lavish songs and dances shot in exotic places such as the Swiss Alps, and dramatic action packed with synthetic and processed sound effects dislocated even further from the sonic reality of otherwise vibrant Indian locations and landscapes. Largely studio-centric and industry-dependent technicians tended to construct a film's sound environment by artificial means, typically paying little attention to auditory authenticity and using songs and louder background music as a way of masking the sound design's shortcomings. In some sense, this trend approached sonic over-modulation, manipulation, and abstraction in enhancing sound's emotional and affective qualities, playing on the fringes of the audience's imaginings and fancies, such as by the processing of the voice of the antagonistic character or bodily sound effects of a side character and so on. For example, the specific sonic representations of villainous characters were constructed using vocal manipulation as well as extended reverb on their footsteps and other bodily postures and gestures, producing visceral responses in audiences by affective mimicry (Plantinga, 2009, p. 94), leading to popular mass appeal in films like *Sholay* (Sippy, 1975), *Dharmatma* (Khan, 1975) and *Coolie* (Desai, 1983).

These post-synchronization practices of the then popular mainstream Indian cinema created several aesthetic problems, however. Most relevant to my analysis is the lack of actual spatial information in the recorded and designed sounds rendered by this technique, as well as the inability of this style of sound organization to provide evidence of the situated realities in the Indian film's reconstructed pro-filmic spaces (Lastra, 2000).[17] Actors had to perform twice: once on location in front of the camera, and once more in the studio in front of the studio microphone, where real situations and sounds on location would be impossible to re-create. Most of the actual

17 As introduced earlier, 'pro-filmic space' is defined by film sound scholar James Lastra (2000) as the space of the fictitious site in front of the camera, a space that is reconstructed in the film space for producing the cinematic experience by recording, layering, and designing sound.

location sound recordings were replaced in the studio with processed Foley and canned sound effects used to mask the technique's shortcomings. These effects ended up sounding inordinately loud and high-pitched, while locative sounds were often neglected.

Digital Recording, Sync Sound, and Surround Design

Since the late 1990s, a large-scale conversion from analogue recording and analogue production practices to digital technologies was taking place in Indian cinema.[18] Digital technology was integrated into the production and post-production stages of filmmaking, as well as in the reproduction and projection formats. The ramifications of Indian cinema adapting to digital technologies have been far-reaching and are particularly evident in the way filmmaking has changed through the novel practices of digital sound production (Kerins, 2011; Holman, 2002). Production practices and techniques such as location-based multi-track sync sound recording[19] and surround sound spatialization have altered the notion of the film soundtrack[20] in the digital realm of Indian cinema since the 2000s. This process of digitalization has had a substantial impact on the narrative strategies and aesthetic choices made with cinematic sound production, informing the creation of the cinematic space and the presence of the site[21] in the pro-filmic space by novel digital techniques such as sync sound and surround sound design that contributed to a new sense of diegesis involving more mimetic elements. With this digital shift, the mise-en-sonore or the

18 See: interview with Aloke Dey and Anup Mukherjee in *Between the Headphones: Listening to the Sound Practitioner* (Chattopadhyay, 2021b).

19 As already introduced in the introduction, 'sync sound' is an abbreviation of synchronized sound recording made on film location or the film set, gathering mostly the live performance of actors' voices, effects, and locational ambience, often using newer technologies of digital multi-track Hard Disc recording, which can store large audio data that can be later utilized in surround sound spatialization techniques.

20 As discussed earlier, the term 'soundtrack' is widely debated in cinematic sound studies largely due to its usage, denoting a linear optical track on the filmstrip mixed with an accompanying music track, thus transmitting a sense of linearity and one-dimensionality. I have argued (Chattopadhyay, 2013, 2015, 2016, 2017, 2021a) that 'soundtrack' can be a limiting term in cinematic sound studies; shaped by the methodologies of the digital realm, sound in cinema transcends the linear representation of a fixed 'track' and moves towards an elaborate and fluid spatial environment.

21 'Place' is a generic term, 'site' is more specific. For my analysis, I will use 'site' more often than 'place' to specify the narrative depiction of specific locations in Indian cinema.

auditory setting[22] has also been reconfigured. The auditory setting produced after the digital revolution in filmmaking contrasts considerably with earlier cinematic experiences born in the monaural and stereophonic frameworks of sound production. Both of these latter frameworks were inclined to take a screen-centric storytelling strategy, while the advent of digital sync sound and surround design resulted in more off-screen spaces. Therefore, it is necessary to consider these transformations considering the aesthetic choices, strategies, and novel spatial experiences they have sparked in order to reach a thorough understanding of the implications of digital technologies on modes of sound production in Indian cinema. I consider Indian cinema's intrinsic changes following the digital revolution in the 2000s and the ensuing technological transformations in recording, track-laying, spatialization, mixing, and sound reproduction. This comparative study helps reveal the implications of digital technology on sound production within a broader historical trajectory.

Contemporary Indian cinema in the digital realm facilitates deliberate sound practices to create cinematic experiences that are spatially 'present' (Lombard and Ditton, 1997; Skalski and Whitbred, 2010; Grimshaw, 2011). This mode of sound production differs considerably from earlier production practices.[23] This study reveals how digital technologies, such as multi-track sync recording and surround sound design, impact the organization of sound to produce more situated cinematic experiences. Later, I will use significant examples from post-2000 Indian cinema, which is not only the world's largest producer of films (Times of India, 2013), but also a vibrant market for digital film technologies such as Dolby Digital, DTS and Dolby

22 The term 'film space' is defined as the space that the spectator encounters, a space that is organized and constructed (e.g., the linking of shots through sound editing and sound design). On the other hand, the area in front of the camera and sound device's recording field is known as the 'pro-filmic space', as discussed earlier in this chapter. Combining these two definitions, it can be argued that the choice and arrangement of pro-filmic space substantially affect the spatial dynamics of the mise-en-scène of sound or, if I may take liberty of using an unofficial but useful coinage, 'mise-en-sonore' or the *auditory setting* – the actual sonorous environment, spatial organization of ambient sounds, that the listener experiences – a setting that in turn influences the verisimilitude or believability of a film in the ears of the audience member. For further discussion, see Chattopadhyay, B. (2021a). *The Auditory Setting: Environmental Sounds in Film and Media Arts.*

23 The analogue monophonic sound production era conveyed certain observed and recorded evidence of the fictional site through synchronized means with an aesthetic of realism, and later the dubbing era induced the somewhat remote and site-unspecific conditions in the auditory setting via dubbing and sound processing, as well as the non-inclusion of ambient sounds into the scheme of sound organization.

Atmos.[24] Drawing on existing works in film sound production (Kerins, 2011; Sergi 2004; Holman, 2002; Sonnenschein, 2001), the contemporary practice of sound in Indian cinema is unpacked, as elaborated in the introduction, in light of the narrative strategies of diegesis and mimesis. Considering the notions of presence, current practice is dissected to critically listen to how spatially evocative sonic environments are constructed as opposed to the spectacular, interruptive song-and-dance sonic rituals typically found in Indian cinema (Gopalan, 2002; Rajadhyaksha, 2009). The creative and innovative sound practices of the digital era are leading to a new realm in both the independent Indian scene as well as established popular cinema. Audiences can increasingly connect with the film experience through spatial perception (Waller and Nadel, 2013) and auditory cognition (McAdams and Bigand, 1993). Crafted with sync recording and surround sound design of multiple sound layers recorded from actual locations, these practices can provide audiences with an embodied experience of sound. This can be termed a 'cinematic soundscape' if we consider the notion of soundscape (Schafer, 1994; Drever, 2002) as a point of departure for studying a spatial evocation of sound in cinema. We can consider this as a shift away from the linear and spatially-limiting notion of the film soundtrack. These significant shifts in production practices emphasize the need for a critical listening-driven approach when studying sound in contemporary Indian cinema.

Digital multi-track 'sync'[25] recording and multichannel surround sound mixing offer a wider palette of sound material for designing a spatially elaborate mise-en-sonore in cinema. With the advent of digital technology, widely available and easy-to-handle recording devices, applications, and facilities have made various options and formats available to sound practitioners.[26] Scholars of sound production Tomlinson Holman (2002) and Mark Kerins (2006, 2011) inform that digital sound systems (DSS) have introduced a number of possibilities, including significantly larger dynamic ranges of over 100 dB, a fourfold improvement over its monophonic predecessor and almost double that of the stereophonic format (Holman, 2002; Kerins, 2011); a larger headroom[27] of 20 dB (a major improvement to

24 See: www.dolby.com/in/en/index.html

25 Throughout this book, I use this widely used term in film industry, 'sync,' which, as explained earlier, is an abbreviation of synchronized sound recording made at the location of shooting.

26 See: interview with Anup Deb in *Between the Headphones: Listening to the Sound Practitioner* (Chattopadhyay, 2021b).

27 'Headroom' means the amplitude above a designated reference level that a sound signal can handle before it distorts or 'clips'.

the 6 and 12 dB headroom of monophonic sound); six discreet channels (5.1 surround sound) and more in other multi-channel formats such as Atmos, wider panning for sound spatialization; and full-frequency channels (20 Hz–20 KHz) with a flatter response. These capacities have made new production practices possible: a wider range of dynamics in sound, such as low frequency rumbles to whistles, as well as increased complexity in mixing and spatial fidelity (Kerins, 2011), while recording and processing the available depth, perspective, and width[28] of sounds collected on location and brought into the digital audio workstation for cleaning and track-laying.

In Indian cinema, the on-location sync sound recording technique has been a direct result of this trend of digital innovation in sound production, which means that sounds are recorded on location in synchronization with the image during shooting. In this widely used term in the industry – 'sync' – the emphasis is on the 'synchronization' aspect of recording sound, pointing to the fact that the practice qualitatively differs from earlier post-synchronized dubbing, therefore proliferating a completely different set of narrative methodologies in approaching the pro-filmic space and creating the spatial fidelity in the cinematic story-world. Interviews with prominent Indian film sound practitioners (Chattopadhyay, 2021b) also give insights into these inherent transformations and their ramifications in the narrative diegesis. In this practice, multiple options for organizing numerous tracks for location sounds or the guide tracks, sync effects, dialogue, and background musical scores open up possibilities for recording a larger number of sound elements in multi-track formats. The increased storage space of digital formats also allows for recording and mixing additional ambient sounds after shooting in order to capture the intricate details of a location. These extensive recordings are incorporated in post-production stages without the need to reuse sparse archival content from stock sounds and pre-recorded ambience. In the studio there are a variety of digital applications to manipulate recorded sounds to restructure their site-based characteristics to fit the cinematic narrative. As the new trend of sync sound and surround design became widely accepted in contemporary Indian films, sound production incorporated newly available digital technological innovations over the existing set-up. Post-production techniques – editing, designing, and mixing in the studio – became faster and the projection of sound in theatres and multiplexes moved toward

28 See: interview with Aloke Dey in *Between the Headphones: Listening to the Sound Practitioner* (Chattopadhyay, 2021b).

multichannel surround systems such as Dolby 7.1, Auro 3D, and Dolby Atmos.[29]

What is aesthetically different in this new era of digital technology-dominated film sound production as compared to the earlier techniques of optical synchronous 'direct' sound recording, monaural mixing, magnetic recording, dubbing, and stereophonic re-mixing? Film scholar Gianluca Sergi (2004) asserts that the early digital surround sound mixing practice relied on 'the same screen-centric notion of cinema sound as their mono and Dolby stereo predecessors' (quoted in Kerins, 2011, p. 5). But he also pointed out 'a reassessment of the relationship between screen sound and surround sound' (quoted in Kerins, 2011, p. 5) in the later technological innovations of surround sound. These statements suggest that surround sound technology shifted the preconceived idea of screen-centric sound (mono as well as stereophonic) towards a wider area of diffused sounds surrounding the cinema screen. Film scholar Vivian Sobchack expresses this changeover as 'shifts of emphasis and attention in both sound technology and our sensorium' (2005, p. 2), leading towards what Rick Altman has termed 'greater realism' (1992, p. 159), predicting the future of sound production in cinema in terms of technological innovations that support realistic representations of place, as I have elaborately discussed in *The Auditory Setting* (2021a). In the contexts of Indian cinema, these technological and associated aesthetic transformations were observed in greater intensity than its American counterparts as pre-digital sound production was indeed marred by a mediocre to inferior quality of sound recording and mixing.

I have shown in the first subchapter that the emergence of any new technology in Indian cinema generates a great deal of discussion and deliberation about its potential use or abuse, as the advent of colour did to black and white films or sound recording did to silent Indian films – but there has always been great enthusiasm and readiness in utilizing a new technology. The question of how 'stereo' should sound has been much debated since the advent of stereophonic sound in the 1950s, when cinematic sound was already standardized in accordance with the monophonic recording-production-reproduction chain, from direct sound recording to its projection in monaural theatres in India. In the context of Hollywood,

29 The few early Indian films to be released in the Dolby Atmos format were *Sivaji 3D* (Shankar, 2012) and *Madras Café* (Sircar, 2013). Amid these developments, Dolby faces a rival in Auro 3D, which arrived in Indian cinemas with *Vishwaroopam* (Hasan, 2013). Both companies develop and offer audio technologies that digitalize, split, and route sounds into multiple channels of surrounding speakers.

Gianluca Sergi has described this transition as a change from a low-quality optical monaural soundtrack to a relatively cleaner (via tape-based magnetic recording) and better-quality Dolby stereo soundtrack with higher dynamic range, wider depth, and signal-to-noise ratio, but with problems in sound localization still at the fore. He raises questions about the contribution of stereo to cinema by stating:

> This design (i.e., early Dolby stereo) follows the principle that audiences should be offered directional sound (i.e., sound whose direction could easily be identifiable) only from one wall of the auditorium, namely that where the screen is placed. The notion at the core of this thinking is that sound emanating from somewhere other than an onscreen source would cause the audience to get distracted in an attempt to locate the origin of that sound, hence disrupting the narrative flow. Thus, the implied suggestion is that the surround channel be employed only in a diffuse, non-directional manner so as not to 'disturb' the narrative. Despite implicitly suggesting that primary information ought to originate from the screen, the one-wall principle did away with the need to deal with complicated alternatives, like additional surround channels, that would have meant a serious rethink of the meaning of stereo in the cinema. (2004, pp. 20–21)

From what Sergi writes, it is evident that more channels of sound meant a rethinking and reordering of the existing set-up (monaural and stereo mix) in order to achieve a new spatial organization of sound in cinema. If we refer explicitly to practical and experiential accounts from Hollywood, sound designer of note David Sonnenschein refers in relation to his own practice to the addition of channels to the existing normative structures of routing and mixing sounds in order to design different elements of the soundtrack for emerging surround sound design:

> In the LCRS[30] (Dolby SR and Ultra-stereo) system, the dialogue normally projects from the centre with effects and music coming from the left, right, and surround speakers. Ambiance and music can take advantage of the multiple sources to create a space within which the audience can be enveloped. ...With the addition of other speakers beyond the basic four LCRS, the variables increase and more discrete placement can be made with the sounds. (2001, p. 47)

30 Abbreviation for left, centre, right, and surround channels.

Sonnenschein suggests that stereophonic cinema makes it possible to create an extra off-screen space allowing the audience to engage with the directionality of sound. As I discuss in-depth in *The Auditory Setting* (2021a), this capacity opens up a new spatial orientation towards the fictional space, creating the possibility of a dynamic sonic experience in which the sound can move around and beyond the screen (Sergi, 2004) in a relatively more flexible manner. Mark Kerins claims that the stereophonic system had its own limitations in the off-screen diffusion of sound (2006, p. 43) but in the digital surround era, possibilities of forming off-screen spaces are ample. Likewise, in the new digital surround sound environments, the 'off-screen space' has now been expanded with added channels in the front, sides and rear directions as well as on top (e.g., Dolby Atmos), which emphasize a spatially evocative sound environment instead of offering a linear one-dimensional 'soundtrack', with voice, effects and background music mixed into a single track (i.e., in earlier monaural mixes). Kerins argues that in comparison to the screen-centric monaural and stereophonic soundtrack, digital surround systems 'spread out into the theatres as their makers see fit' (2006, p. 43), granting the sound practitioner more creative freedom in the narrative strategy of organizing sound components. Kerins also argues for the 'spatial fidelity' that is provided by digital surround systems. The sound practitioner uses digital systems to put forward 'more perceptible sounds' in the surround channels 'to build multi-channel environments. They assume that audiences will understand sounds originating in the surround channels to be part of the same diegetic space as those originating on screen' (2011, p. 70), spatially expanding and substantially enriching the spatial sound environments.

I mention in *The Auditory Setting* (2021a) that film sound scholar Giorgio Biancorosso also placed emphasis on the spatial reordering of sound in order to create diegetic space in cinema, with the shift from mono to stereo and digital surround:

> After all, sounds whose sources remain unseen not only reach us at all times, but are also crucial in guiding our sense of inhabiting a certain kind of space, specifying its properties and suggesting the kinds of activities taking place therein. It is fair to assume that we bring this ability to perceive the space around us through sound to bear on the construction of a diegetic space. Digital Surround Sound depends on it. (2009, p. 263)

Likewise, the growing digitalization of filmmaking in India allows for the appreciation and construction of a diegetic space to which previous sound practices, in mono and stereophonic frameworks, paid lesser attention. There is an increased audience demand for authenticity due to a proliferation of news channels, live broadcasts of contemporary issues, and more public discourse on social justice, as well as the ubiquitous reality shows on television. This thrust for credibility and authenticity also in cinematic diegesis leads to more site-specific spatial details to be included in an embodied experience of sound. The spatial organization makes the audience convinced of the presence of the site within the narrative construction of the story-world, which, as demonstrated in this study, is often achieved using sounds in the multi-channel sonic environment of contemporary Indian films aspiring to present the realities of today more 'from the ground' than previously.

Apart from merely adding more audio channels to the digital surround environment, the digital era of Indian cinema engenders an emergent fascination with urgent social issues, authentic people, marginal voices, real locations over sets, and more detailed and accurate documentary evidence from socially active sites in India – noticeable in production practices such as sync sound in Indian films – which suggests a rediscovery of cinema's origins in cinematic realism.[31] For example, in recent mainstream Indian films, the preceding practices of dubbing, stock sound effects, and studio Foley are gradually being replaced by authentic, site-specific sync sounds (Chattopadhyay, 2012a). These sound layers incorporate a wider dissemination of naturalistic and site-specific auditory artefacts into the creation of believable film spaces, adding depth, texture, and perspective. This reordering of film spaces has been gaining momentum with the increasingly direct participation of sound technicians in the filmmaking process through their involvement in location-based sync recording, production mixing, and surround sound design. Digital multi-track sync sound has been accepted as a highly precise, artistically demanding and skilled recording technique practiced by sound technicians, involving the use of actors' original dialogue, thereby eliminating tedious post-production processes such as dubbing and Foley in current Indian films. This book endeavours to unpack such tendencies in sound practice and sonic aesthetics.

31 Referring to the direct sound recordings made in Indian cinema between the 1930s and 1950s.

References

Altman, R. (ed.) (1992). *Sound Theory/Sound Practice*. New York: Routledge.

ArZan (2019, October 14). "Ardeshir Irani, the Father of Indian Talkies who had many other Milestones to his Name". *Parsi Khabar*. https://parsikhabar.net/film/ardeshir-irani-the-father-of-indian-talkies-who-had-many-other-milestones-to-his-name/21171/ (Accessed 13 April 2023).

Bhatt, O. (2019). "Writing About Sound: The Early Talkie Film Periodicals of India". In *International Perspectives on Publishing Platforms* (pp. 45–81). London: Routledge.

Bhatyacharya, T. (ed.) (2009). *Dhanimoyotar Itibritto*. Kolkata: Roopkala Kendro.

Biancorosso, G. (2009). "Sound". In *The Routledge Companion to Philosophy and Film*, edited by Carl R. Plantinga and Paisley Livingston (pp. 260–267). London: Routledge.

Birtwistle, A. (2010). *Cinesonica: Sounding Film and Video*. Manchester: Manchester University Press.

Bloom, Peter J. (2014). "Sound Theory". In Buckland, W. and Branigan, E. (eds), *The Routledge Encyclopedia of Film Theory* (pp. 430–434). London: Routledge.

Burch, N. (1985). "On the Structural Use of Sound". In E. Weis and J. Belton (eds), *Film Sound: Theory and Practice* (pp. 200–209). New York: Columbia University Press.

Chattopadhyay, B. (2012a). "Sonification of Cinema: Studying the Use of Location Sound in Indian Films" (conference paper). *Music & The Moving Image Conference*, NYU.

Chattopadhyay, B. (2013d). "The Cinematic Soundscape: Conceptualising the use of sound in Indian films". *SoundEffects* 2(2), 66–78.

Chattopadhyay, B. (2015). "The Auditory Spectacle: Designing Sound for the 'Dubbing Era' of Indian Cinema". *The New Soundtrack* 5(1), 55–68. Edinburgh: Edinburgh University Press.

Chattopadhyay, B. (2016). "Being There: Evocation of the Site in Contemporary Indian Cinema". *Journal of Sonic Studies* 12. https://www.researchcatalogue.net/view/286592/286593 (accessed 10 September 2023).

Chattopadhyay, B. (2021a). *The Auditory Setting: Environmental Sounds in Film and Media Arts*. Edinburgh: Edinburgh University Press.

Chattopadhyay, B. (2021b). *Between the Headphones: Listening to the Practitioner*. Newcastle upon Tyne: Cambridge Scholars Publishing.

Chion, M. (1994). *Audio-Vision: Sound on Screen*. Translated and edited by C. Gorbman. New York: Columbia University Press.

Deming Jr., W. (1995). "Talking Pictures in India". In Maitra, Prabodh (ed.), *100 years of Cinema*. Calcutta: Nandan.

Doane, M. A. (1985). "Ideology and the Practice of Sound Editing and Mixing". In E. Weis & J. Belton (eds), *Film Sound: Theory and Practice* (pp. 54–62). New York: Columbia University Press.

Drever, L. J. (2002). "Soundscape Composition: The Convergence of Ethnography and Acousmatic Music". *Organised Sound* 7(1), 21–27.

Holman, T. (2002). *Sound for Film and Television*. Boston: Focal Press.

Grimshaw, M. (2011). *Game Sound Technology and Player Interaction: Concepts and Developments*. Hershey: Information Science Reference, IGI Global.

Kuhn, A. and Westwell, G. (2014) *A Dictionary of Film Studies*. In Oxford Reference Online: www.oxfordreference.com/view/10.1093/acref/9780199587261.001.0001/acref-9780199587261-e-0659# (accessed 1 March 2023).

Kerins, M. (2011). *Beyond Dolby (Stereo): Cinema in the Digital Sound Age*. Bloomington: Indiana University Press.

Kerins, M. (2006). "Narration in the Cinema of Digital Sound". *The Velvet Light Trap* 58 (Fall), 41–54.

Lastra, J. (2000). *Sound Technology and the American Cinema: Perception, Representation, Modernity*. New York: Columbia University Press.

Lombard, M. and Ditton, T. (2006). "At the Heart of It All: The Concept of Presence". *Journal of Computer-Mediated Communication* 3(2). https://academic.oup.com/jcmc/article/3/2/JCMC321/4080403 (accessed 10 September 2023).

McAdams, S. & Bigand, E. (1993). *Thinking in Sound: The Cognitive Psychology of Human Audition*. New York: Oxford University Press.

Mukherjee, M. (2012). *Aural Films, Oral Cultures: Essays on Cinema from the Early Sound Era*. Jadavpur: Jadavpur University Press.

Plantinga, C. (2009). "Emotion and Affect". In Plantinga, C.R and Livingston, P. (eds) *The Routledge Companion to Philosophy and Film* (pp. 356–365). London: Routledge.

Rajadhyaksha, A. (2009). *Indian Cinema in the Time of Celluloid: From Bollywood to the Emergency*. New Delhi: Indiana University Press.

Rajadhyaksha, A. (2016). *Indian Cinema: A Very Short Introduction*. Oxford: Oxford University Press

Schafer, R. M. (1994). *The Soundscape: Our Sonic Environment and the Tuning of the World*. Rochester: Destiny Books.

Sergi, G. (2004). *The Dolby Era: Film Sound in Contemporary Hollywood*. Manchester: Manchester University Press.

Skalski, P. & Whitbred, R. (2010). "Image versus Sound: A Comparison of Formal Feature Effects on Presence and Video Game Enjoyment". *PsychNology Journal* 8(1), 67–84.

Sonnenschein, D. (2001). *Sound Design: The Expressive Power of Music, Voice, and Sound Effects in Cinema*. Studio City, CA: Michael Wiese Productions.

Sobchack, V. (2005). "When the Ear Dreams: Dolby Digital and the Imagination of Sound". *Film Quarterly* 58(4), 2–15.

Waller, D. A. and Nadel, L., (eds). (2013). *Handbook of Spatial cognition*. Washington DC: American Psychological Association.

3. Direct Sound and Early Talkies

Abstract: Chapter 3 traces the various modes and facets of sound production in Indian cinema during the early talkies utilizing what is often termed 'direct sound'. It delineates the advent of sound in Indian cinema and the ensuing transition from the silent era to sound films, analysing how the complex evolution of sound changed methods of storytelling and narration. Starting with the use of sound in early Indian talkies such as by the *Bombay Talkies* studio (1934–53), the chapter elaborately describes the development of direct sound and the resultant linguistic divisions in Indian cinema, using several monaural films as example. Focus is placed on determining monoaural aesthetics, and audio-visual and audio-audio relationships.

Keywords: Indian cinema, sound studies, early talkies, direct sound, optical recording

The Sonic Leakage

Throughout the trajectory of monaural sound practice in Indian cinema since 1931, available recording techniques and equipment have had a somewhat limited dynamic sound recording and production range, leading to voice and dialogue being at the top of the priority list, as previously discussed. Likewise, the freedom of a microphone in outdoor locations or on a film set had been reduced by controlling its directionality, to focus on recording 'almost always the voice' (Chion, 1994, p. 5), thus establishing a sound-cinema that was of an essentially vococentric nature. It is no surprise then, that the films produced in these times were mostly devotional song and music-oriented mythical or religious stories where the primary narration was carried out through either dialogue between characters or post-recorded voiceovers in some cases. Despite the limitation of the dynamic range, and the controlling and suppression of other sounds such as ambient and incidental sounds, some louder locational environmental sounds may have

Chattopadhyay, B., *Sound in Indian Film and Audiovisual Media: History, Practices, and Production*. Taylor & Francis Group, 2024
DOI 10.5117/9789463724739_CH03

unwantedly intruded into the film's directly recorded soundtrack. The predominant emphasis had been on the voice of the actors, but these stray and fundamentally unwanted sounds could provide spatial information about the real site that the film was shot in. In a well-known sequence from the legendary Indian film *Devdas* (Barua, 1935/1936), the eponymous protagonist Devdas is languishing over his initial arrival at a brothel in Calcutta (now Kolkata) following his recent breakup with Paro, a childhood sweetheart from his native village. Devdas' fateful interaction with the 'prostitute' Chandra leaves him in a state of perpetual melancholia, claustrophobia, and remorse from which he can never recover. The mise-en-scène indicates that the story is taking place in an indoor location within a closed building in a busy city in Calcutta, but the incidental sound of a bird call appears and continues throughout the entire sequence alongside the actors' directly recorded voices and an abundance of background musical score. A sensitive listener may locate the bird call that continues throughout the entire sequence and disappears with a cut. The significance of this sound element lies in the direct recording of sound from the very location where the film was shot: Calcutta. That off-screen sound of a bird call in *Devdas* frames the distinct realistic evidence of pro-filmic space and the diegetic world mimetically captured without the practitioner's knowledge in film space during the direct sound practices in early Indian talkies. What this intrusive sonic leakage hints at is the existence of a vibrant world that was otherwise not allowed to enter into the cinematic universe – not because of technological limitations but because the story-world was preferably kept insular from the lived world. In the opening sequence of *Daktar* ('Doctor', Majumdar, 1940) there is a suggestion of what this lived world would sound like as the running train driven by the enthusiastic vibrations of the engine advances towards a new city and suddenly encounters a calf standing on the railway line. The train decreases its speed and halts just near the calf, while a number of local boys rush to the site and remove the lovelorn calf from the rail. The entire episode resonates with the sound of the train engine, the atmosphere of a rustic village landscape, and the murmuring chatter of the local boys. However, as soon as the calf is removed, all these sounds are replaced by a song sung live by the protagonist Pankaj Mullick, who happens to be on the train heading to the new city. In the audio-audio relationship, a hierarchy where the song is given far more screen time and space over the spatial information about the cinematic site is maintained. However, the little sonic details that leak into the story leave the listener aware of the sonic reality surrounding the protagonist (e.g., the railway line through a village), before his voice takes over the entire screen.

In the typical Indian film shootings from this time, the camera establishes a shot and the sound recording device follows it in order to capture a limited sound field within the visual frame, displaying attention to the available sound-producing objects in accordance with the mise-en-scène and the story-world. In most cases, the director and the cameraperson would determine the microphone placements. Within the given space and time of shot-taking, there was a scattering of different sources of sound. Those that weren't related to the script of the film's storyline were considered unwanted noise. Shyam Benegal rightly points this out in an interview I conducted:

> For instance, if you think of *Alam Ara* and so on, you will find that they actually had sync sound. That was the old methodology. They used to have these studio cameras – the big, very unwieldy things and they were properly sound insulated. So, the machine sound couldn't be heard. They used to have mics and the sound recordist usually used to sit at the corner of the recording studio and you had to keep absolute silence. The studios that were made for silent cinema had to be, again, insulated. Despite that, once in a while you would have pigeons coming in and I have experience of that. (Chattopadhyay, 2021b, p. 20)

As mentioned in earlier research (2021a, 2015), in a vococentric script the freedom of a microphone is reduced, as its directionality is forced to focus on recording 'almost always the voice' (Chion, 1994, p. 5). Within the limited scope of direct recording, available sound sources are narrowed to a minimum on the recording media. Elsewhere in his interview, Benegal informs about the direct recording method: 'That was the methodology of recording. They used to record on 35 mm' (Chattopadhyay, 2021b, p. 20). In spite of the limitation and suppression of the dynamic range, these stray sound elements like pigeons, birds, traffic, and natural and other incidental or site-specific sounds intrude into the film's predetermined soundtrack and may turn out to be capable of carrying meaning about the nature of the locale and making the film space lively. The off-screen sound of a bird call (along with its many other off-screen sounds such as a passing car horn, a barking dog, or the clamor of playing children in other sequences) in *Devdas* (Barua, 1935) holds distinct documentary evidence of the cinematic site in the film space by direct sound recording represented on the film's soundtrack as aurally 'realistic' (Kania, 2009, p. 244) in perception, even if the sound is incidental, off-screen, off-script, and not a deliberate usage. These sounds unintentionally entered onto the scripted film space through a microphone boom that couldn't be closed like a pair of ears. Indian film scholar Madhuja Mukherjee terms this the 'here and now' effect created by such a register:

> In early talkies, where sound is in sync, on screen, and mostly diegetic, the resonances and meanings are somewhat different from the theories of 'disembodied' sound and music destroying the 'aura' of performances. ...Certainly, these are mechanically recorded images and sounds, which have been recorded from multiple positions and camera angles, in multiple spaces, and thereafter have been edited and restructured. Moreover, as we listen (and see) we first hear a mechanical sound, then a voice, words, rendition, and the sound of music. Nevertheless, it can recreate the 'aura of performance' in its own terms as the star /actor (who is also a singer) sings in a time which is real, where the real and reel time become one. In many cases the mechanical rendition of the song is a continuous take, and has a strong 'here and now' effect. (2007, pp. 52–53)

Even with a predominant mode of control and musical masking of a natural auditory setting, Indian films of this time exemplify the occasional recording of incidental sounds because of the direct recording technique used on location and the actors performing live in front of the camera. Later, in the works of Satyajit Ray, we will see that this 'here and now' effect is made evident and explored in the narrative strategy of a 'direct mono sound' aesthetic whereby the presence of a lived world becomes prominent in the deliberate use of sounds recorded from respective locations at the time of shooting.

On the qualitative difference between talkies and films that explored sound, René Clair states: 'The talking film is not everything, there is also the sound film' (1929). What made the 'sound films' different from the talkies was the complex layer of sounds besides 'almost always voice' (Chion, 1994) and the background musical score. These layers of sound included sync effects and, most importantly, the site-specific atmospheric sound, or the 'ambient sounds' – noted categorically by early sound practitioners (Chattopadhyay, 2021b) – as the crucial elements of creative intervention for creating a sense of place and spatiality despite being often thought of as unwanted noise that shouldn't enter recording media.

Aural Situated-ness

Achhut Kannya ('Untouchable Maiden', Osten, 1936) ends with a climactic journey to a fair and a festival ground – a journey that turns out to be quite decisive in terms of the fate of the titular character Kasturi, a Dalit girl. The film deals with the social position of the Dalits in India, and through its dramatic elements leading to the tragic climax, develops a reformist approach to the deep-rooted social problem in India around caste. Most parts of the last sequences in this film are shot on location as is apparent from the sound quality and the intensive presence of a very convincing sound of a cart, and the incidental and other environmental sounds throughout the journey and at the fair site.

The journey begins with Pratap, the Brahmin boy who had grown up with Kasturi, and deeply loves her, but is now unhappily married to a Brahmin girl from his own caste following the societal norm of his village and his mother's insistence. Pratap is driving a bullock cart towards the fairground to sell some of the unsold merchandise from his family shop; his father was banned from selling his wares due to his friendship and proximity to Kasturi's father, a Dalit. As the cart enters a busy village scene, the resounding squeaking and creaking of the cart is greeted with the local sonic environment that is vibrant and alive with a multitude of cattle and man-made sounds, passersby and their chatter, and murmers of the forest. These are the sounds, which are not usually allowed to enter the film space and the Indian mise-en-sonore. However, in this dialogue-less outdoor sequence, these sounds creatively anchor the story-world to the reality of the everyday quotidian life of an Indian village. These directly recorded sounds provide aural situated-ness, making the narrative development credible. As I discussed in the introductory chapter, this sense of aural situated-ness and spatial realism works as a system of narrative processes through a balanced combination of diegesis and mimesis, drawing upon the plausibility of the film space vis-à-vis the physical world as lived and experienced. The following longer sequence takes place at the fairground: in this longer sequence without dialogue, a typical fairground location is revealed through ambient and incidental sounds. Diegetic music becomes part of the ambience consisting of different musical instruments played in the open-air fair ground, the gathering of people, man-made sounds and rhythmic instruments played out in the open-air theatre, excited shouts and murmurs of people speaking, wind instruments playing to invite visitors, whirligig and amusements ringing, and hawkers screaming to sell their products. All these sounds are distinct, but together they facilitate place-making, situating the audience in the crowd through spatial fidelity from directly recorded monaural sound.

Achhut Kannya was a production of Bombay Talkies which was founded by the maverick producer and director Himanshu Rai and the gifted actor

Devika Rani; the latter played the main character of Kasturi in the film. In the 1930s and 1940s, Bombay Talkies was known for being an innovative and technically resourceful studio. The Malad-based studio was well equipped with film and sound technologies imported from Europe, and had sound and echo-proof stages, film development laboratories, optical editing rooms, and even a preview theatre. To maintain its international standards and reputation, the studio invited experienced European film practitioners such as German film director Franz Osten and German cinematographer Josef Wirsching to work in India. Their contribution, as film historian Rahaab Allana notes, 'brought a European internationalism to Indian cinema and aspects of modernism also took root in the secular themes of films like *Achhut Kanya* that questioned untouchability' (Allana quoted in Tilak, 2017). No wonder this social realist commitment as well as aesthetic standard and technical expertise was reflected in their productions. Take for example *Izzat* (Osten, 1937), *Prem Kahani* (Osten, 1937), and *Kismet* (Osten, 1943).

In *Prem Kahani* there is a sequence in which the protagonist Jagat visits Bombay and his father's old friend Bhagwandas comes to pick him up at the station. They then embark on a car ride through the city towards home. The discussion between them is surrounded by the city sounds and the machine-like ambience of the car's interior. The clarity of their directly recorded voices along with the realistic reverberations inside a moving car and, most importantly, the city outside the open window buzzing with busy quotidian sounds are all captured and produced with professionalism. *Izzat* is shot mostly outdoors and the quality of the outdoor recordings once again proves the technical expertise that the production studio had. The auditory settings of *Kismet,* on the other hand, remain mostly indoors or feature nightly outings through the city as the heroine cannot move due to an accident and the hero primarily works as a thief. The indoor spaces resonate with clear and naturalistic room reverb, situating the story inside the colonial houses of Bombay with their high ceilings and corridors.

What about other contemporary productions in India besides the revered production house and studio Bombay Talkies at that time? There were indeed a number of production houses with substantial outputs like Rajkamal Kalamandir, Prakash Pictures Studios, Filmistan, Mehboob Studio, Sagar Movietone, Wadia Movietone, R. K. Films and Studio, Minerva Movietone, and the Navketan Films to name but a few.

Founded by film producer Vijay Bhatt, Prakash Pictures Studios produced many popular films of this period, ranging from religious and historical films to family dramas. For instance, the film *Station Master* (Luhar, 1942) from the production house can be considered an entertaining piece of comedic family drama situated in the everyday reality of Indian middle-class households, familial settings, and their aspirations. The film starts at the office of the Rangpur railway station's stationmaster. The sound of the recurring telegraph introduces an inert and busy signal office with the chaotic sounds of a buzzing platform and station area bleeding inside this order. We get to know the protagonist right away: a thoughtful, responsible, and conscientious stationmaster, who takes care of his post. Getting up from his chair, he comes out onto the platform and is greeted by a cacophony of sounds that range from hawkers and tea vendors to murmurs of waiting passengers. With this multitude of incidental and ambient sounds, the atmosphere of a typical Indian railway station is established in the film space, situating the audience in the auditory setting of the station and the lives of the railway workers.

Vocal Synchronization

Early talkies were all direct sounds in terms of recording the voice, which meant that the voice was registered then and there while the dialogues were performed in-situ. With the direct recordings on location or on set, the voice was automatically synchronized with the characters shown and heard in the cinema. This changed as soon as ADR (Automatic Dialogue Replacement, more popularly known as 'dubbing' as explained earlier) was incorporated into Indian cinema roughly from the late 1940s. I will discuss the effect of dubbing in a later chapter. In this chapter I shall critically listen to some of the early talkies where lip sync or vocal synchronization led to the credibility of the story.

As I mentioned earlier, many major sequences of *Izzat* (Osten, 1937) were shot outdoors and the quality of the outdoor recordings of the dialogue carries the naturalistic lack of reverb usually found in an open space or field. It was a Hindi-language social drama produced by Bombay Talkies. Set in a rural environment in central India, the story delineates the struggle between two tribes: the powerful and rich *Marathas* and the relatively poorer and weaker tribal or indigenous community known as the *Bheels*. In the narrative, there has been lasting peace in the region for many years due to the leaders of both communities having agreed a peace pact. However, as soon as the heir of the *Maratha* leadership changes, the new leader, Balaji, nephew of the older and outgoing leader, starts a self-serving policy of animosity towards the *Bheels* and eventually sets out to displace and dispossess them. Initially the *Bheels* restrain themselves from reacting to

Balaji's insinuations because of the prevailing peace treaty and the overall friendly attitude of most of the *Maratha* people, but, after the murder of one of their fellow men and on the orders of the new leader, they decide to put up resistance. Much of the screen presence of their leader Kanhaiya, played by Ashok Kumar, takes place outdoors, as he absconds and takes a guerilla approach to counter the new *Maratha* leadership. His speeches, given in open fields, stand out from the theatricality of the dialogue-throw of early Indian talkies. Other dialogues including those of his lover Radha, played by Devika Rani, are all reassuringly in sync with the moving lips. This merging of the lip movements and corresponding voice became rare with the emergence of dubbing in Indian cinema in the late 1940s. I have shown in chapter 2 that since the introduction of synchronized sound in Indian cinema, monaural sound practice from recording to mixing stages and to reproduction meant that 'a single channel of sound was played from a loudspeaker placed behind the screen creating the illusion that the sound of the film was emanating from the projected images' (Kuhn and Westwell, 2014). In the case of the voice, it was the lips that convincingly appeared as the source of emanation of the sound, making the dialogue believable to the audience.

As many critics have pointed out, Ashok Kumar started a more natural style of acting and dialoguing compared to the prevailing styles of theatricality in the early talkies. From his first screen appearance in *Jeevan Naiya* (Osten, 1936), Kumar used the frankness of his directly recorded synchronized voice as an instrument for screen presence. He learnt from contemporary Hollywood films that acting was not merely saying one's dialogue but listening to others around and to the setting. According to Director Tapan Sinha, 'He is the man who showed that film acting is something else. He began to speak and to behave normally.' This style of voicing the dialogue, evident in *Izzat* (Osten, 1937) and his other works, was in synchronization not only with the moving image and the visuals but also with the setting, since he was being context-aware in his live performance. This sense of synchronization with the time and space helped move Indian cinema from the crass theatrics and *Jatra*-style dialogue in earlier films to a cinematic language marked by quotidian sensibility, grounded appeal, and reliability through the lip-sync voice.

Regarding this mode of lip-sync, sound practitioner Amala Popuri notes, 'The importance of sync, lip sync, is that it puts you in orientation with the visual. In the absence of that you are a little disoriented' (2021b, p. 516). Likewise, films made with vocal synchronization – direct sound recording of voice on the optical media, for example – are far more credible, grounded, and relatable in terms of the way listeners connect a body as the source of a voice in everyday conversations. In the dubbed films that were later mass-produced in India, this lack of voice-body synchronicity created an uncanny disconnect as most sound practitioners in the collection of interviews note (Chattopadhyay, 2021b). In a sequence from *Prem Kahani* (Osten, 1937), the conversations between two characters inside a running car are interrupted by the traffic sounds outside and there are sudden jerks in the voice when the car goes through some bumps on the road. The audience can register the authenticity in synchronized voice recording with all its rough edges.

This sense of sonic authenticity is something subsequent social realist filmmakers explore and imbibe as a narrative strategy. In an interview I conducted, Shyam Benegal notes:

> I refused to take a pilot track because at that time I was kind of fanatical about getting the right and exact sound. I was very much impressed with the idea, which was very current in Europe at that time – the use of direct sound. The whole idea was that you had a sense of reality that you got with it (direct sound) – impossible to get in the post-sync – the emotion, the performance. The performance was not broken into two parts – the visual performance and the sound performance. It was an integrated performance. (2021b, p. 22)

The credibility issue of the voice is paramount when a filmmaker aims to communicate the message to a sensitive and receptive audience. *Jeevan Naiya* (Osten, 1936) was a popular film because the direct recording of the grainy and imperfect nasal tone of Ashok Kumar resonated with the quotidian believability of the audience members. What Roland Barthes terms 'the grain of the voice' is that texture and tonality of the voice that strives towards a connecting sound without a semantic mode of realizing the everyday relational world (1977, p. 182) in which non-humans speak with humans through sonic gestures and auditory touch, the known vibration of a person close to us (through such grains we understand beyond the semantic) and connect to their presence. Direct sound vocal synchronization could achieve an authentic connection, which dubbed voice could never achieve – there was always a disjuncture. The long telephone conversations between Kumar and the heroine Devika Rani are engaging, as their intimacy was easier to establish because of the vocal synchronization with speaking lips and the sound of their voices matching in tempo, unison and in spirit, mixed on location with the spatial information of the room with all the reflections and possible noises around. In close up, this matching was more audible. In the school sequence where Devika Rani interacts with a bunch of children, their cumulative noise was not processed out but kept as directly recorded information. In the monologue sequence of Devika Rani towards the end, the room reverb was natural in the way it was recorded along with the grain of the voice with occasional hesitations and breathing in between – all recorded then and there, adding to a resounding sense of believability.

Music-on-the-go

Devdas (Barua, 1935) opens with an apparently non-diegetic background score followed by the song *Balam Aye* ('Come, my Beloved'). But eventually, one of the primary characters Paro, played by Jamuna Barua, hears the song on her way through a forest. She tries to locate the source of the song, as she knows by the voice that it's sung by her beloved Devdas, played by K.L. Saigal. She listens around and follows the trails of the sound in order to find him. Yet, the song remains a non-diegetic soundtrack until the next sequence in which we see Devdas sitting under a tree singing that song with his eyes closed, and there is no other musician in sight. A curious and sensitive listener may ask where the musical instruments are. Are they hidden in the forest? How is it possible that we see the source of the voice, but the accompanying musical instruments remain invisible? How is it possible that part of the song is diegetic, and the rest is not? These are valid questions. And in the case of most pre-digital Indian films where song and saccharine/sweet background music are in abundance, such questions are even more pertinent. But these questions are rarely asked either by the cinema-going public or critics. All these cinematic aberrations are taken with a pinch of salt for mass entertainment. In the case of the direct sound era's production of songs, we learn from veteran practitioners such as Jyoti Chatterjee and filmmakers such as Shyam Benegal that musical instruments in early Indian sound films were indeed hidden in bushes and behind walls before playback systems were invented. Benegal exposes the process of recording songs and music on the go in an interview:

> But one of the innovative practices of the time, particularly with Prabhat, was that they actually recorded songs outdoors. If you see some of those films – if you go to FTII[1] you should have a look at some of the films made

1 The Film and Television Institute of India, Pune, is one of India's two premier state-sponsored film schools. The other one is SRFTI, the Satyajit Ray Film and Television Institute, Kolkata.

> in the 1930s in Prabhat studios – they used to have a sound recordist with a boom mic. When it was being shot, the sound recordist would be recording sitting on a *thela* (hand drawn cart) and the instruments also had to be playing. You couldn't have any elaborate kind of group. So you had to have a harmonium and a table – two/three instruments – which were carried along. You had the camera on a trolley in front, and then you had this *thela* for the music to be played, which was being recorded simultaneously as the girl or boy, whoever, was singing. So this used to be done. That was the system. There was no playback during those days. That's how it started. It was very interesting. But it certainly restricted the acting style because the people couldn't take their heads away from where the microphones were. (2021b, p. 20)

As we see and hear K.L. Saigal singing underneath a tree while a mid-long shot is being taken, the musical accompaniment is not made part of this diegetic world. The musical instruments are placed at a safe distance. Veteran sound mixing specialist Anup Mukherjee states that a single microphone was used in such music-on-the-go sequences to record the songs at the location along with the character's voice (Chattopadhyay, 2021b). This process of synchronizing on site came back in the sync sound era. In the direct era, however, during the live recording of songs on location the question was, 'where will the *tabla* player sit?' The *tabla* could be put on top of a tree, or the harmonium could be placed behind the tree. So, keeping in mind the proper balancing of this, the recordists used to fix the spot of their seating where the musician wouldn't be visible, and the song would also be aesthetically pleasing to hear in terms of its live production. This technique evolved within a few years because the songs were recorded on fixed cameras. The music-on-the-go situation was like this according to Mukherjee: 'a group of musicians are playing flute and they are standing inside a small pond neck-deep in water and playing the flute' (Chattopadhyay, 2021b, p. 102) – just to avoid the camera's viewfinder. As a seasoned sound practitioner, Mukherjee thinks, 'What I feel for is the hardship they had done during that time and a change that happened since the time the playback system came' (Chattopadhyay, 2021b, p. 102). I will discuss the aftermath of the advent of playback systems in the next chapter. But at this point, I would like to underline the use of quasi-diegetic music in Indian films prior to the arrival of playback systems to cater to the expectation of a smooth entertainment experience. Benegal points out how this mode of quasi-diegesis was problematic but worth trying:

> So, it was somewhat tied up in...they were kind of imprisoned because of the placement of the mic and they had to be facing the mic all the time. The demand on the actor was quite enormous. Firstly, the person had to sing, had to be in tune, instruments were to be played in tune. While doing all this, even the slightest shifting this way or that way meant that the recording would go bad. We had such kind of problems, but it was very fascinating and very interesting. (Chattopadhyay, 2021b, p. 21)

Many interesting but often fantastical and unrealistic sonic experiments can be found in the direct-era Indian films based on mythological or fantasy themes like *Lal-e-Yaman* (Wadia, 1933), *Alif Laila* (Bhatt and Dave, 1933) and *Hatimtai Part I-IV* (Sethi, 1933), most of which were derived or adapted from Parsi Theatre – a popular form of staged plays in Mumbai and most of western India. Film scholar Sharmistha Gooptu suggests that Indo-Persian and Islamic cultures have had a major influence on Parsi Theatre (2010). Along the lines of *The One Thousand and One Nights* (or Arabian Nights), Parsi Theatre performed adventure-romances drawn from Parsi mythological stories rather than focusing on the contemporaneous social realities of Mumbai or India; these plays were later adapted into early Bollywood films to capitalize on the popularity and mass appeal they had at that time. K. Moti Gokulsing, and Wimal Dissanayake in their book *Indian Popular Cinema: A Narrative of Cultural Change* explain this trend as 'blended realism and fantasy, music and dance, narrative and spectacle, earthy dialogue and ingenuity of stage presentation, integrating them into a dramatic discourse of melodrama' (2004, p. 98). To trigger their mass appeal and sales, these films incorporated humour, melodious songs, background music and dance sequences added with sensationalism and exotic stagecraft, in which the sounds were often estranged from the realistic setting and rendered exotic, and the music used was traditional tunes with accompanying orchestras hidden from the built film sets.

More social realist but popular films like *Prem Kahani* (Osten, 1937) produced by Bombay Talkies on the other hand, did make minimal modulations to their sounds and music compared to their mythology-focused counterparts. Khorshed Minocher-Homji (later known as Saraswati Devi), one of the few early female composers of Indian cinema, composed music for this film along with many others of the period, much of which were based on Indian classical music and the *Khayal* genre. In some of these live music settings, Saraswati Devi herself sang some of her songs, making diegetic interventions in the films as characters. These songs were soulful and reflected the moods of the protagonists. As we will see, these social realist films were precursors to the Golden Age.

References

Barthes, R. (1977). *Image, Music, Text.* Translated by Stephen Heath. London: Fontana.

Chattopadhyay, B. (2015). "The Auditory Spectacle: Designing Sound for the 'Dubbing Era' of Indian cinema". *The New Soundtrack* 5(1), 55–68. Edinburgh: Edinburgh University Press.

Chattopadhyay, B. (2021a). *The Auditory Setting: Environmental Sounds in Film and Media Arts.* Edinburgh: Edinburgh University Press.

Chattopadhyay, B. (2021b). *Between the Headphones: Listening to the Practitioner* (2021b). Newcastle upon Tyne: Cambridge Scholars Publishing.

Chion, M. (1994). *Audio-vision: Sound on Screen.* Translated and edited by C. Gorbman. New York: Columbia University Press.

Clair, R. (1929/1985). "The Art of Sound". In Weis, E. & Belton, J. (eds), *Film Sound: Theory and Practice* (pp. 92–94). New York: Columbia University Press.

Gokulsing, K. M. and Dissanayake, W. (2004). *Indian Popular Cinema: A Narrative of Cultural Change.* London: Trentham Books.

Kania, A. (2009). "Realism", in P. Livingston and C. Plantinga (eds), *The Routledge Companion to Philosophy and Film* (pp. 237–247). London and New York: Routledge.

Mukherjee, M. (2007). "Early Indian Talkies: Voice, Performance and Aura". *Journal of the Moving Image* 6, 39–61. Kolkata: Jadavpur University Press.

Tilak, Sudha G. (2017, December 27). "A German Cinematographer's Love Affair with Indian Cinema". BBC News Services. www.bbc.com/news/world-asia-india-42422599 (accessed 1 March 2023).

4. Sound of the Golden Age

Abstract: The fourth chapter studies sound in the so-called 'Golden Age' of Indian cinema, when the quality and quantity of production skyrocketed and reached a global stage. From the late 1940s to the late 1960s, a resurgent period of film production – not limited to the Hindi-language films of Bombay but spread throughout the Indian subcontinent after independence from colonial rule in 1947 – galvanized a legitimate national cinema that would later be called 'Indian Cinema'. The era's films demonstrated quality and quantity in equal measure. They resonated with the hope, enthusiasm, and collective commitment to nation-building based on a newfound sense of freedom and self-determination, liberated from the spectre of oppressive colonial censorship.

Keywords: Indian cinema, film history, sound studies, Golden Age, social resurgence

Social Resurgence

Hindi cinema's so-called Golden Age, from the late 1940s to the late 1960s, was a resurgent period of film production across the Indian subcontinent after independence from colonial rule in 1947. The era's films possessed quality and quantity in equal measure. They resonated with the hope, enthusiasm, and collective commitment to nation building based on the newfound sense of freedom that had beaten the spectre of oppressive colonial censorship. Many of the film directors and practitioners who pushed the industry towards its new heights were dedicated followers of India's anti-colonial movements and struggle for independence. The emergence of a modern, post-colonial India with nationalist aspirations galvanized their work, despite the harsh reality of grappling with a dwindling economy after 150 years of British exploitation. The fervor bled into regional film industries, which centred on local language film-works with distinctive provincial character. The technological means and sonic methods adapted

Chattopadhyay, B., *Sound in Indian Film and Audiovisual Media: History, Practices, and Production.* Taylor & Francis Group, 2024
DOI 10.5117/9789463724739_CH04

in regional and state cinemas across the country were often similar to those used in Bombay's large-scale industry, making the case for a legitimately subcontinent-wide national cinema that could be called 'Indian Cinema'.

Several popular Hindi-language classics such as *Mother India* (Khan, 1957), *Pyaasa* ('Thirsty', Dutt, 1957), *Kaagaz Ke Phool* ('Paper Flowers', Dutt, 1959), *Awara* ('The Vagabond', Kapoor, 1951), *Shree 420* ('Mr. 420', Kapoor, 1955), and *Mughal-E-Azam* ('The Great Mughal', Asif, 1960) come from this period. Global lists, forums, festivals, and screenings regularly featured some of these films as representative of Indian cinema. Indeed, these films influenced the development of a distinctive cinematic language on the subcontinent for decades to come.

As mainstream, Bombay-based Hindi films started to flourish, parallel cinema cultures, focusing on narrating India's social realities, emerged to challenge the formulaic storytelling of more economically successful films. Though India's alternative cinema found its own zenith later in the twentieth century, and more recently through independent, digitally distributed films, works from the Golden Age such as *Neecha Nagar* (Anand, 1946) and *Do Bigha Zamin* (Roy, 1953) laid the ground for alternative sensibilities to be sparked and encouraged. Parallel regional cinema also resulted in new Bengali, Tamil, and Malayalam cinematic lexicons. Out of all of India's critically acclaimed regional cinemas, Bengali cinema in particular saw its own golden era unfold in the early 1950s with interventions from stalwarts like Satyajit Ray, Mrinal Sen, and Ritwik Ghatak, alongside the likes of Tapan Sinha, Arundhati Devi, and Ajay Kar – some of Bengal's most celebrated filmmakers. They ventured into filmmaking and made their formative works during the Golden Age but developed their own independent styles and subjects into the 1970s and 1980s.

When India gained independence, cinema's rebirth as a mass communication tool and social medium in the nation-building project gave the industry impetus to move away from India's 1940s mainstream mythological and fantasy-based escapism, and cinematic versions of theatre pieces and social melodrama. Many highly successful films probed contemporaneous social realities and themes about the new state, the working class, and the chaos of a resurgent, urban middle-class life. Critically acclaimed films made prior to this turning point rarely found commercial success, while films with wider appeal were mostly considered too stylistically normative and industry-driven to merit critical appraisal. It wasn't until the 1950s that several films, including *Awaara* (Kapoor, 1951), *Pyassa* (Dutt, 1957), *Mother India* (Khan, 1957), and *Mughal-E-Azam* (Asif, 1960), found both commercial success and somewhat appealed to film critics in India and abroad due to their renewed commitment to the cinematic craft, subject matter, and content, as well as their approaches to storytelling. *Mother India* (Khan, 1957) was a highly successful commercial film, while also winning awards nationally and internationally: the film received India's first Academy Award nomination. A number of films from this Golden Era found wider appreciation on the European film festival circuit: for example, *Neecha Nagar* (Anand, 1946) was awarded the Palme d'Or at Cannes Film Festival.

Vococentric Sound

Many of the late talkies were made with a transitional methodology from direct sound to voice replacement techniques using early forms of dubbing. This approach to voice and the mediatory device of dubbing framed the clean voice as a diegetic storytelling tool. The qualitative approach to film production in the formative years of India's national cinema paved the way for the Golden Age. The sense of aural situated-ness in many of the major sequences in talkies made at the end of the 1950s were based on the reconstruction of auditory settings during re-recording stages. Re-recording is understood as a process of controlling and registering disparate film sound components on a single strip of optical track in parallel with the moving image track as a so-called 'married print'. As discussed in the Introduction, aural situated-ness emerges from a sense of presence determined by how sounds are recorded from their site and reconstructed in the film space from their locative sources. In *Neecha Nagar* (Anand, 1946) one locates certain outdoor scenes in an open field adjacent the 'lowly city' from the sounds of birds and cattle. But the field is in the background, its sound almost absent. The heated discussion in the citizen's assembly is the focus. Their voices carry the primary narrative: the revolt of citizens from the poorer part of the city against Sarkar, the oligarch, and the aftermath of the divide and rule policy. In a later indoor sequence, this vococentric storytelling takes centrestage, as Sarkar, in a clean voice recording, describes *Neecha Nagar*'s context and its urban architecture, including the city's planned gentrification model alongside replanned irrigation. I argue that the description of the site would be more evocative, grounded, and relatable with the creative use of site-specific environmental sounds, but these sound components remain under-utilized. This vococentric mode of film sound practice during the Golden Age, where voices carry the narrative forward, became central to mainstream Indian Cinema.

Taxi Driver (Anand, 1954) was shot on location in Mumbai with many of the city's sites enlivened by embodied sounds from a cab engine. Chetan

Anand is said to have carried a microphone in a car running across the city. Although the voices of the film's characters and their expressive songs dominate, on the few occasions of relative verbal silence, the city resonates with its urban clamor and human activity. In the classic *Naya Daur* ('The New Era', Chopra, 1957), the relatively careful organization and re-recording of sounds, including train horns, engine murmurs, tanga bells, the babble of speech, and horses trotting, establish the railway station's location in a small village near Bhopal. In a crucial sequence, we hear the conflict between tradition and modernity in verbose exchanges. As the film's protagonist, Shankar, a *tangewala* (cart puller), meets his love Rajani for the first time, while she is waiting for a tanga to take her home. The son of the local landlord, Kundan, also arrives but from the big city with the intention of modernizing the village and local businesses by investing in technologies, machines, and automation. Kundan plans to introduce a bus to the village, which would take away the tangewalas' livelihood. The sequence starts with a train entering the village station, sparking the enthusiastic roar of *tangewalas* inviting passengers to their carts. As the horse-driven carts start to move against the sound of the train engine, the human-machine conundrum needs resolution. As the story develops, the various stakes in the story manifest through a series of misunderstandings, leading to a climactic race between a local *tanga* and an imported bus. In winning the race, the desperate Shankar, concerned with the abuse of technologies in the hands of profiteers, declares his commitment to reinstate village traditions despite the pressure of modernization in independent India.

In the popular comedy drama *Mr. & Mrs. '55* (Dutt, 1955), the escape of an estranged couple from urban neurosis to a village location helps soothe their nerves, facilitated by the sound of birds chirping, the breeze blowing through tall trees, children laughing, and the grounding sounds from domestic spaces such as a hand-turned wheat grinder. However, this sylvan and gently inhabited soundscape is interrupted by passionate exchanges between the lovers and an obtrusive song.

These few examples show how ingrained character dialogue and songs were in Indian cinematic storytelling. Now, let's take a closer look at what mainstream Indian cinema sounded like before and during the Golden Age. If we pick a few pertinent examples from the 1950s and 1960s – popular, regional Bengali romantic comedies starring Uttam Kumar and Suchitra Sen, or those in Hindi starring Raj Kapoor among other stars – we hear mostly vococentric representations of sound, which provoke a mode of listening that is primarily 'out-of-place'. By coining 'out-of-place', I mean the lack of spatial information in the diegesis, marking a relative absence of the site depicted in the narration. The monophonic arrangement of sounds mainly consists of actors' voices and a few synchronized sound effects that remain suspended in the abundance of music; we hardly relate to the sites in the construction of the pro-filmic space. Site-specific ambient sounds are absent in this sound strategy, rendering urban and other locations implausibly silent and relatively disembodied within the primacy of dialogue. All we hear is continuous talking accompanied by an overly melodramatic background musical score, sporadically punctuated by often unintelligibly arranged song sequences intended to entertain.

However, a parallel thread of Indian cinema existed in which sound was approached creatively as a purveyor of social realism. In a latter chapter, I discuss the work of Satyajit Ray. Both Ritwik Ghatak and Mrinal Sen, whose work will be outlined later in another chapter in detail, also started working independently during this period of Indian cinema. How do we posit these filmmakers in the auditory context of the Golden Age? How do we study their creative sound practice within the mainstream practices that became standardized in Hindi, and by regional film industry giants like the Bengali cinema of Kolkata and Tamil film industry in Chennai? Indian film scholar Sharmistha Gooptu sheds light on Ray's position in this context and his larger contributions to developing a realistic sensibility in sound practice that underlines the importance of location, space, and spatiality developed creatively through a nunaced practice of sound:

> Bengali film practice was no longer the same post-Ray. While there were earlier films like *Chhinnamul*, which was shot on location and was starkly realist, or directors like Bimal Roy who had worked with rank newcomers, it was with Ray that such practices became more purposefully institutionalized within the Bengali film industry. As Marie Seton notes, 'Between the release of *Pather Panchali* and *Apur Sansar*, Ray's films exerted one notable influence upon Bengali films, and to a lesser degree on Hindi films. Other directors were made aware of the value of location recording'. (Gooptu 2011, p.165)

Likewise, we hear relatively carefully organized sonic layers, including spatial and site-specific sounds and locative sound effects in some sequences, in many Bengali popular classics such as *Suno Baranari* ('Listen O Lady', Kar, 1960) and *Chaowa Pawa* ('To Want and to Receive', Yatrik, 1959). The long passage of railway transfers in the former and the elaborate platform sequences in the latter are adequate evidence of a realistic portrayal of place and site incorporating a spatial practice of sound in the monaural organization of recorded sound in cinema, making a social realist ethos audible.

But, on the other hand, cinematic locations remain sonically impoverished in many popular mainstream films, because ambience wasn't considered an important layer of the monaural soundtrack. The urban sites in several popular Bengali films such as *Deya Neya* ('Give and Take', Bandyopadhyay, 1963) and *Chowringhee* (Mukherjee, 1968) are rendered silent from the marked absence of site-specific sounds, and careful sound recording and organization. In the opening sequence of Mukherjee's *Chowringhee*, the protagonist Shankar meets an old friend at the busy crossing of the Esplanade (Chowringhee area of central Calcutta). Although a lot of traffic is visible on screen, we only hear the actors' dialogue. Social realist films, in contrast, treated the same busy crossing differently: in Ray's *Mahanagar* ('The Big City', 1963) and Sen's *Calcutta Trilogy* (1971, 1972, and 1973) the location receives elaborate sonic detail, creating an immediate, direct, and real sense of presence of downtown Calcutta through a layered narrative methodology of organizing diegetic sound elements carefully listened to on site.

Playback System

The Golden Age is known for the rise in popularity of film songs. Initially, actors sung the songs themselves and their singing ability was an advantage for getting work. However, it was rare to find actors who could both act and sing well. A new breed of singers was therefore needed who could efficiently lip-sync with actors on screen. Film songs found greater success in subsequent decades, witnessing a prolific period of film music around the 1960s and 1970s with renowned composers such as Sachin Dev Burman, Anil Biswas, Khayyam, and Naushad at their peak. The Golden Age also saw some of the most well-known voices in film songs such as Mohammed Rafi, Kishore Kumar, and Lata Mangeshkar emerge due to a conducive industry and playback singing methods. The era celebrated film songs in particular, laying the foundation for music and songs ornamenting and punctuating the narrative lines in popular Indian cinema.

I have already discussed how actors used to deliver their dialogue directly into the camera between 1931 and the late 1930s. They also used to sing on location, while their songs were recorded directly onto the filmstrip, as with dialogue. Dubbing was yet to be used and hadn't yet taken a foothold in the production process. As discussed, for the production team, direct sound had a few problems related to incidental noise, because studios were equipped with primitive sound and camera technologies that created noise on site. Often a single camera was available to shoot the entire film, which was heavy and difficult to move around. If the characters moved within the location or on the film floor while singing as part of their shot action, the direct recording would be noisy due to location crowd, traffic, and other incidental sounds. This problem was solved with the invention of the playback system. Veteran sound practitioner Anup Mukherjee illuminates on this history in *Between the Headphones: Listening to the Practitioner*:

> The playback was more or less parceled from India because in Hollywood playback wasn't popular in that manner. The songs were recorded separately the way orchestra is generally recorded and then it was shot. Once the playback system started, it reduced the dependency quite a bit. The camera became much smoother and moving at the same time, which helped or added in shooting various moments of the songs separately. So the playback system was one where the song was recorded prior and it was played back while shooting and the synchronized performance of the camera along with the playback without which it won't synchronize. ...So, according to the playback of the song the artist would give the lead of the song, the song playback would happen, and it would be shot from different positions and compositions. (Chattopadhyay, 2021b, pp.102–103)

This early form of playback was first used in the Bengali film *Bhagya Chakra* ('Wheel of Fate', Bose, 1935a) and its Hindi remake, *Dhoop Chhaon* ('Sun and Shade', Bose, 1935b). The central idea was that if the song was recorded first on the film strip and then played back for the camera taking the shot as 'song picturization' with the acting characters lip-synching the song, then the quality of the recorded sound would be better in terms of noise reduction and clarity. This innovative method was successfully developed by Nitin Bose's sound technician brother, Mukul Bose, who assisted music director Raichand Boral, using the voices of singers that performed in the film, namely K.C. Dey, Parul Ghosh, and Suprabha Sarkar. Because the recorded song was played back during the actual shooting and recording of other sound effects, this method became known as the playback system, which would rule film music production for years to come.

Irrespective of the success of the playback system to replace music-on-the-go during on-location shooting, many drawbacks and mistakes occurred. There are numerous examples where a singer giving the voice to an on-screen actor faltered miserably. In the Bengali film *Pratham Basanta* ('First Spring', Gupta, 1971), the song continues after the character closes her mouth. Often, the singer's voice lacks verisimilitude with the actor's tone and grain. Even when voices do match, the songs often sound out of corporal presence; it was only the benevolence of audience ears that absolved these incongruities for their listening pleasure.

Synchresis is a curious terminology coined by French composer and film sound theorist Michel Chion in *Audio-Vision*, which describes 'the spontaneous and irresistible weld produced between a particular auditory phenomenon and visual phenomenon when they occur at the same time' (1994, p. 63). In the playback system, the synchresis of lip movement on screen and a separate body singing the song created a precarious convergence in the listening mind of the audience. This precariousness manifested in the way a playback singer became successful in singing for a particular actor, their joint stardom created through postproduction collaboration. Indian stars like Dev Anand collaborated frequently with singer Kishore Kumar, Raj Kapoor with singer Mukesh, and Uttam Kumar with singer Hemanta Mukherjee. Similar to dubbing and ADR, such convergence of the separated voice and an on-screen body was, however, often suspended between acceptance and disbelief. The chemistry between sung voice and spoken voice was often a crucial factor in this regard. When the lips moved and the sung voice fell through due to faulty synchronization, it was only the audience who would be lenient and accept it, giving the song the benefit of the doubt.

Direct Dubbing as a Realist Technique

According to Chion, there are moments when the background music in films seems to elevate on-screen action through synchresis. The process differs from synchronization: while the sound and vision may move in separate directions, they meet in a few fertile and key moments to emphasize the narrative action and its prevalent mood. They don't always have to match, and this freedom of movement and convergence in both vision (e.g., lip movement on screen) and sound (e.g., a played-back song or dubbed voice) allows a reasonable amount of creativity for the sound re-recordist and mixing engineer – and now the sound designer in the digital era – to create a dynamic audiovisual relationship. However, without the careful organization of sound, involving the practitioner's artistry and capacity to attune nuances, this convergence may go awry. In the case of Indian cinema, as previously discussed, populist commercial films made in the dubbing era increasingly faltered in achieving synchresis through a lack of nuance.

Social realism-oriented films limited playback singing and dubbing for this very reason. Their filmmakers and sound technicians looked for innovative methods intended to provide spatial verisimilitude that manifested not only in the elaborate use of site-specific sounds but also in the recording and re-presentation of voices. In an interview, Ray's sound mixer Jyoti Chatterjee speaks about the director's use of 'straight or direct dubbing' to avoid some of the camera and production noises. It is a method to capture the actor's voice immediately after the shot is taken – unlike the standard loop dubbing used in Indian cinema from the 1960s onwards.[1] 'Direct or straight dubbing' is done in the same place and under similar circumstances to retain the spatial authenticity of the site and the sited performance of the actors, something that couldn't be recreated in the studio in front of the looped reference of visual images. We find evidence of such innovative dubbing in Ray's films *Nayak* ('Hero', 1966) and *Aranyer Din Ratri* ('Days and Nights in the Forest', 1970). This practice of random dubbing on location (to retain the site's details and ambience, when direct synchronized recording confronts logistic hindrance) shows a staunchly realist filmmaker like Ray's commitment to creating a presence of the site in diegesis, expanding the possibilities of forming a cinematic reality as close to the experientially real, site-aware, and lived sonic environments as possible.

Auditory Balance

Importantly, the audio-audio relationship in most Golden Age films was well balanced. As discussed in the Introduction, the complex relationship between different sound components – namely, voice, music, sound effects, and ambience, or environmental sounds – needs to be handled well to offer the audience a balanced sonic experience. An auditory equilibrium in film suggests an acoustic ecology within its diegetic universe. There are typical hierarchies between these components, and often one component (e.g., voice) is given more importance over others (e.g., ambience) in narrating a story or depicting a situation.

If we listen carefully to some of the films produced during the Golden Age, for example *Dharti Ke Lal* ('Children of the Earth', Abbas, 1946), *Shair* ('Poet', Chawla, 1949), *Mother India* (Khan, 1957), *Pyaasa* ('Thirsty', Dutt, 1957), *Kaagaz Ke Phool* (Dutt, 1959), *Awara* ('The Vagabond', Kapoor, 1951),

1 See: interview with Jyoti Chatterjee in *Between the Headphones: Listening to the Practitioner* (Chattopadhyay, 2021b).

Shree 420 (Kapoor, 1955), *Mughal-E-Azam* ('The Great Mughal', Asif, 1960), and *Gunga Jumna* (Bose, 1961), their sound pallets were quite carefully handled when incidental or ambient sound layers were incorporated, and attention was paid to location-specific details in a sequence with less focus on the voice. For example, most outdoor scenes in these films included an optimal amount of traffic sounds and other incidental sounds present on location. This balance can be understood as an outcome of the first standardization of technologies and sonic methods used during the Golden Age, which was then destabilized when dubbing took over from the 1970s onwards, focusing on the voice, sound effects, and loud music.

References

Chattopadhyay, B. (2021b). *Between the Headphones: Listening to the Practitioner* (2021b). Newcastle upon Tyne: Cambridge Scholars Publishing.

Chion, M. (1994). *Audio-vision: Sound on Screen*. Translated and edited by C. Gorbman New York: Columbia University Press.

Gooptu, S. (2011). *Bengali Cinema: An Other Nation*. London: Routledge.

5. Tracing the Sound in Satyajit Ray's Film-works

Abstract: This chapter develops a careful study of the works of filmmaker Satyajit Ray. His work with sound is analysed, conceptualizing the term 'audiographic realism' that emphasizes the actual site-specific sound elements recorded and incorporated in the monaural rendering. Ray is considered an auteur and cultural icon – his legacy being felt on the fringes of mainstream Indian cinema. However, the specifics of sound practice and the nature of sonic experience in his films remain largely underexplored. This chapter examines how Ray's treatment of sound recognizes aural situated-ness and actuality through the keen observation of place and locative listening, using direct sound recording in monaural and synchronized practice, which created a precedency for realistic auditory settings now championed by filmmakers in the digital era.

Keywords: film sound, ambient sound, monaural aesthetics, Satyajit Ray, synchronized sound, audiographic realism

Sounds of a Transformative World within the Home

One wonders about the ramifications of the disruptive intrusion caused by the jarring sound of an electric generator in renowned Indian filmmaker Satyajit Ray's (1921–92) award-winning[1] film *Jalsaghar* ('The Music Room', 1958),[2] appearing in an otherwise lyrical and nostalgic auditory setting. In this sequence, the landowner-protagonist Biswambhar Roy is enjoying a *Hindustani* classical *Vina* recital by a musician in his music room while the faint sound of a motor enters the room, gradually becoming prominent.

1 All India Certificate of Merit for the Second-Best Feature Film (1959) and National Film Award for Second Best Feature Film in Bengali (1959).

2 An excerpt from the film: www.youtube.com/watch?v=jbxgVjlFZBo.

Chattopadhyay, B., *Sound in Indian Film and Audiovisual Media: History, Practices, and Production.* Taylor & Francis Group, 2024
DOI 10.5117/9789463724739_CH05

The landowner asks his porter, 'What is the sound I hear?' with a slight furrowing of his eyebrows. He is obviously disturbed. The porter replies that it is the sound of a generator coming from the house of his *nouveau riche* neighbour Mahim Ganguly who has recently got electricity. The music is paused while the landowner ascends to the rooftop and the new-fangled sound of the generator engulfs the landscape around his dark and desolate palace. Symptoms of modernity thus appear at the cost of disturbing and destabilizing the archaic ecology of sound at the protagonist's home, suggesting a new sonic reordering. We later hear the recurring sound of the generator in a number of sequences and with each occurrence, the contemporary and modern world of rapid and tumultuous change casts a darker shadow over the traditional coziness of the home. However, it is not simply a question of conflict between tradition and modernity that Satyajit Ray feels compelled to articulate, as most Ray scholars argue (Ganguly, 2001; Cooper, 2000), but one can also infer that Ray is concerned with providing a detailed observation and inclusive framing of the worldly elements that reflect societal change in his contemporaneous India through a specific use of site-specific sound in the narrative strategy. Further, these sonic elements are offered in the narration to exhibit the diegetic world as a means to establish the realistic presence of a vibrant outdoors in film space. His films embody an awareness of the existence of an eventful and transforming India within the cozy setting of a tranquil home confined and constrained by traditional forces.

In the use of sound in Ray's films and other Indian films born out of the Parallel Movement (led by the works of Ritwik Ghatak, Mrinal Sen, Tapan Sinha, Adoor Gopalakrishnan, G. Aravindan, and Shyam Benegal, among others), Ray and his comrades in their own inimitable voices are interested in establishing and carrying out a mode of social realism of different degrees and shades that enabled them to speak about the grassroot realities of socio-political conditions, failings, injustices, and the trauma of an independent India after partition, trying to address issues like corruption, poverty, and inequality, challenging background music and song-drenched commercial Indian cinema. I will discuss the Parallel Cinema movement in the next chapter, but here I want to focus squarely on Satyajit Ray's work – with sound as a reference point – given its importance in Ray's films.

Focusing on Ray's film oeuvre, one may position the tendency towards detailed observation of the locations and spatial situations depicted in the narration, providing documentary evidence of the presence of corporeal sites/ people and a growing concern about the realistic social-political conditions of his time throughout his entire body of films – from *Pather Panchali* ('Song of the Little Road', 1955) to *Agantuk* (1992) – with a distinct commitment to realism in the use of sound. He achieves this mode of social realism through a contemporaneous sound practice of analogue synchronized recording and monaural mixing of sound derived from thoroughly observed and alert listening to the sonic environments available in the locations of the films' narrative settings. Such use of sound in some of Ray's representative films is examined here in order to trace out significant tendencies and predilections that suggest an important benchmark and reference point for studying sound practice in Indian cinema through its early developments and future trajectories. His contribution to the evolution of cinematic sound in India is thereby situated in relation to the Parallel Cinema movement and later in lending his legacy to Indian independent cinema trends in the digital era. I intend to show how in Ray's films 'we perceive the sound not only in temporal sync, but also in correct spatial placement, as our brains create the bridge to reestablish a normality to the situation' (Sonnenschein, 2001, p. 47) adding to the realistic framework of filmmaking and film sound production in the context of Indian cinema (not to mention global cinema as well). It is my contention that Ray chooses to use the specific layers of site-specific sounds and off-screen ambiences among other components such as the direct recordings of characters' voices, incorporating them in the strategy of storytelling in order to produce spatial sensations and the sonic presence of the locative realities in the film space constructed in his body of work.

The sound in most of Ray's films was made primarily through direct recording and analogue single-channel synchronized sound rendered for monaural mixing, as well as monophonic reproduction.[3] I consider that his use of sound generally highlights a distinct recognition of and attention to the pro-filmic spaces depicted on screen with a synchronized sound strategy while still endeavouring to expand the diegetic space. This is done by creating a realistic mise-en-sonore or auditory setting by deliberately making detailed sonic information available in terms of the sounds recorded on site. This approach was a departure from the practice of sound in popular mainstream Indian cinema of his period which relied heavily on the dialogue of characters, background music, and intermittent song sequences to carry out the storytelling while ignoring a thorough and careful design of various other creative sound elements. In this context, Ray's authorial position in

3 See: Ray's filmography as director, www.imdb.com/name/nm0006249/#director; for further information refer to interview conducted with Ray's sound mixing engineer Jyoti Chatterjee in *Between the Headphones: Listening to the Practitioner* (Chattopadhyay, 2021b).

Indian cinema can be understood as prominent and crucial in drawing the attention of future filmmakers to the spatial details and the construction of film space in the use of sound.

We need to consider that Ray worked within the constraints of the monaural sound set-up of Indian cinema. Although stereophonic design, mixing, and reproduction arose during his lifetime – in the late 1980s and early 1990s (Chattopadhyay, 2015), Ray preferred to work with the mixing for monaural soundtrack in all of his films.[4] It is common knowledge that he had to contend with budget limitations; most of the stereophonic films made during this time were big budget productions, but Ray depended mostly on local producers with small budgets. However, whether Ray was at all interested in exploring the new medium and the new technology of stereophony is questionable given the kind of sound aesthetic with vivid synchronized sonic details observed in his films that exemplify monophonic rendering of sound at its best. Perhaps if they were made with the technology of stereophonic mixing, his films would sound less precise in terms of the use of sounds to synchronize these sounds and their spatial situations. There is no space to speculate on this issue, although we have his films in our hands to study the innovation and experimentation he was able to make with the limited resources available to him at the time.

Listening to Ray's Auditory Settings

The *ethos* or the philosophy behind Ray's stylistic features involves detailed spatial observation in the use of sound – this approach manifests as 'situated listening' (Lastra 2000) in Ray's early works. These films make the aesthetic choice of an inclusive and layered design of synchronized sound in the diegesis. Ray's debut film, *Pather Panchali* (1955),[5] provides for seemingly accurate representations of different locations from the village Boral in West Bengal, where the film was shot. This was accomplished by creating situations of listening-in-place in the use of synchronized sound centred behind the screen within a monaural framework. The screen-centric single-channel controls the '[s]ound's movements in and through the specifics of location' (LaBelle et al, 2011, p. viii) in a more spatial as well as temporal manner, using site-specific sounds such as the wind through the grasslands,

4 See: interview with Jyoti Chatterjee in *Between the Headphones* (Chattopadhyay, 2021b); and Ray's filmography as director, www.imdb.com/name/nm0006249/#director.

5 A sequence from the film: www.youtube.com/watch?v=sohAteJynhc (accessed 5 July 2022).

the drone of electrical poles beside the railroad, the friction of tree branches in a gentle breeze in the forest, etc. There are ample examples of practices such as situated listening and applying a narrative strategy for attending to the diegetic universe in the three corresponding films from Ray's *The Apu Trilogy* (1955–59) and other films that followed. In *Aparajito* ('The Unvanquished', 1956),[6] we distinctly hear different zones within Benares (or the ancient Indian city of Varanasi, where most of the story takes place) through the ears of protagonist, Apu, following his exploration of the city. The respective cinematic passages are constructed with primarily sonic elements that make use of the depth of field and perspective available on location as detailed site-specific sonic information. In *Charulata* (1964),[7] a period piece, the elaborate use of ambient street sounds – hawkers, vendors, and their aural antics – inform the audience about the secluded and idle neighbourhood in a reconstructed Calcutta of the 1870s. Perhaps the most poignant example of synchronized sound from Ray's films is the matching of the sound of the electric tram's sliding shoe on the trolley pole each time it hits the pantograph with a flash in the title sequence of *Mahanagar* ('The Big City', 1963).[8] To my knowledge, no other Indian filmmaker of his time thought of incorporating such in-depth detail in the use of sound. Such synchronized sound practice in monaural films in relation to the screen-image could be explained by Michel Chion's well-used term 'audiovisual illusion of redundancy' (Chion, 1994, pp. 5–7), but in Ray's case, the sounds don't merely synchronize with the image; rather, they accentuate the visible objects on the screen by expanding the diegetic space of their situation, whereby an 'off-screen space is obviously frequently brought to life' (Burch, 1985, p. 201). This off-screen space suggests the presence of an eventful outside world within which the narration evolves and an awareness of the larger social issues of contemporaneous India including poverty, famine, unemployment, and the erosion of moral values. The manifestation of such a sound practice makes Ray stand apart; Indian cinema at the time illustrated a screen-oriented narrative strategy with vococentric sound production, using typical entertaining tropes of cleaner dialogues interrupted by songs within a 'protected sonic space' (Birtwistle, 2010, p. 57). In this controlled space, many detailed sound elements were considered unwanted noise. In his review of *Jalsaghar* (1958), Douglas Messerli writes about the inclusion of 'noise' – a sound element typically considered unwanted in films (Lastra,

6 A sequence from the film: www.youtube.com/watch?v=hbPg8wCj-yM (accessed 5 July 2022).
7 Link to the film: www.youtube.com/watch?v=4ML1l09v914 (accessed 5 July 2022).
8 Link to the film: www.youtube.com/watch?v=a6xolbhLA2U (accessed 5 July 2022).

2000). He argues that Ray's innovative incorporation of incidental noises intensifies a sense of emplacement. Messerli writes:

> Another kind of noise is heard in the distance, a kind of rhythmic beating that one might almost call music (certainly Cage might have described it as such). When Roy asks his servant about it, he is told that it is the generator atop his neighbour's house, and from that instant on, except for the moments of musical performances, we hear the generator pumping away in the distance, a symbol of the end of a quiet and placid world in a time of movement and dust, as new goods are transported to the neighbour's house along a dirt road. (2011)

As I have already argued at the beginning of this chapter, the incidental sound of a generator, albeit off-screen, provides an elaborate spatial detail that expands the sonic depth of the space and place, as well as the narrative scope of the scene. In regard to this perceived depth in monaural films, Michel Chion discusses the 'mental' synchronization of the off-screen sounds as providing an 'outside' to the field of vision: 'Traditional monaural film presents a strange sensory experience in this regard...If the character is off-screen, we perceive the footsteps as if they are outside the field of vision – an 'outside' that's more mental than physical' (1994, p. 69). Ray's frequent use of off-screen ambient sounds suggests the presence of the unseen yet mentally perceived outside world in the diegesis. This observation resonates with the reading of film scholar Richard Allen who underscores the dynamic relationship between on-screen and off-screen sounds in Ray's films: 'Ray uses diegetic, off-screen sound as an integrated component of the compositional technique of camera movement and staging in depth... that broadens the physical space of the represented scene and may carry expressive importance...keeping with his realist aesthetics' (2009, p. 93). The broadening of physical space is enriched by the inclusive off-screen incorporation of actual locative sounds, enhancing a mental resonance of site. Ray explains in an interview with Pierre Andre Boutang, 'I can use actual sounds creatively to serve the purpose of music and to allow a certain change in mood to be perceptible to the audience' (1989).[9] In another interview on Kolkata TV, Ray maintained that 'one can use actual sounds to suggest moods' (1980).[10] Such perspectives on the use of sound that involve replacing music with actual site-specific sounds to suggest

9 A clip from the interview: www.youtube.com/watch?v=RWS5dlxwZDc (accessed 5 July 2022).

10 Link to the interview: www.youtube.com/watch?v=VfZjxoYSRHQ (accessed 5 July 2022).

certain mood-situations strongly defy the typical tried and tested strategy of utilizing a non-diegetic musical score in the Indian films of his day.

Apart from considering site-specific sounds to be noise, Indian cinema during this period believed in clean recordings of the voice as vehicles for smooth entertainment at the expense of other layers of sound. The absence of spatial detail was compensated with loud background musical scores, as explained above, and was periodically interrupted by song-and-dance performances that suspended the narrative flow with animated punctuations (Gopalan, 2002). The easily attainable method of using the non-diegetic background musical score in every possible situation of emotive engagement was a normative practice in Indian cinema, serving the narrative purpose of elucidating various archetypical, popularly accessible, and implicit moods such as romantic, violent, and sad. Conversely, Ray's use of music, however sparse, tended to be diegetic, underlining the contours of a mood-situation rather than obviously pinpointing it. Further, his use of sound provides elaborate evidence of the spatial positionality of the characters, their auditory situation, and the site where the narrative was based. In the Boutang interview, Ray states:

> In the contemporary films, I use less, less and less music; I can do without music...After all, ideally one should do without music completely. I think, as, because one has an audience in mind, and one is always afraid, that a certain change in mood will not be perceptible unless it is underlined by music, you use music. Ideally, a film ought to be able to do without music. (1989)

Ray's use of even the sparsest musical score was therefore a compromise; ideally, he would use actual sounds recorded from the location to serve the purpose of storytelling with an aurally situated approach that is based on a socio-spatial sensitivity. We can ask where this influence must have come from. Concerning his cinematic influences, Ray would usually mention American films of the 1950s and 60s. However, in an interview with filmmaker Shyam Benegal, who asked him about his use of ambient sounds, Ray cited influences from European cinema that adhered to a sound aesthetic of perceptual fidelity – synchronized sound recorded on location. Shyam Benegal asks:

> Before you came on the scene, in Indian cinema certainly there never used to be what one might call an ambient sound, you know, effects outside of the synchronized effect. Now, you started it. Why did you do that? I mean, what led you to use sound like that?

Ray answers:

> I think that in the use of sound, we were not very original; we were doing what the best European cinema[11] had done, which we admired very much. We didn't really learn from the Indian cinema, we learnt in fact what not to do rather than what to do. So, we had the examples of the best cinema of the West in mind. And with *Pather Panchali* as well as *Aparajito*, there was the question of filling long stretches of silence, because both films have very little dialogue. Naturally, we had to think what to do there, and all sorts of sound effects were used in order basically to fill those long stretches of silence and also at the same time for them to work...creatively, in the sense that they add to the atmosphere; they also suggest things, which are not suggested by the dialogue or by the images. (1985)[12]

Here, Ray talks of adding to the atmosphere of the site while reconstructing the pro-filmic space inside the cinematic universe. For example, *Nayak* ('The Hero', 1966)[13] is one of Ray's most remarkable stylistic experiments, demonstrating his ability to create atmosphere – such as representing the specific zones inside a train – through the synchronized use of sound within a monaural mix. Each of the cubicles of the train within which the threads of narratives and the various intermingling stories of the personas unfold – the berth where the protagonist sleeps, the dining car, the toilet, and the corridor – appears with thorough, specific, and consistent site-specific ambient sounds (e.g., the different carriage vibrations and subtle resonances that help to differentiate the spatial identities of these sites). The sounds come from behind the screen and are synchronized with the images. This auditory setting establishes and expands the diegetic space by realistic means, perhaps in the way Brandon LaBelle describes sound's potential to demarcate a sense of spatiality 'toward a greater understanding of the interconnectedness of space' (2011, p. viii) through its materiality and its manifestation in the listening.

11 Ray was deeply influenced by European auteurist filmmakers like Jean Renoir, whose film *The River* (1951) was directly shot at various locations in Calcutta. Ray volunteered as a film enthusiast on these sites. Later Renoir personally inspired him to make films. See: the biography of Ray (www.satyajitray.org/bio/renoir_meet.htm), and a seminar speech at American Film Institute, in which he mentions Renor's influence on his work (www.youtube.com/watch?v=TdFi1noweog&t=226s), (accessed 1 March 2023).

12 The interview was released by Film Division, see: https://filmsdivision.org/shop/satyajit-ray-2.

13 Link to the film: www.youtube.com/watch?v=pK5hvo8xwso (accessed 5 July 2022).

In a conversation with Samik Bandyopadhyay for India's national television channel *Doordarshan*, Kolkata, Ray elaborates on the creative sound practice in his work:

> And now more and more I use less and less music, because I can use the improved mixing facilities (sic), and I can use [a] more creative soundtrack, whereby one can use actual sounds almost as you use music to suggest moods and things like that…For instance, in a city[-based] story the city itself provides the noise of traffic, and it provides the mood building soundtrack…In *Asani Sanket* one could well have used folk music in abundance, but I preferred to use bird noises; I preferred to use the sound on *Dheki*, you know, that sort of thing, and I preferred to use wind sound, …I used one particular bird – the woodpecker, which I recorded. I was lucky to be able to record one. And it comes at a very crucial point when Moti dies…one can hear the woodpecker, which is a very shrill and rather alarming kind of, rather eerie sort of sound. I felt the use of music would sentimentalize the scene. So, I decided to use this, which was also a realistic sound. (1980)[14]

As Ray explains, the extraneous noises and 'realistic' sounds (or ambiences) in his films replace non-diegetic music with its emotive-escapist overtones (exploited by previous Indian filmmakers) and serve to anchor the listener in the real world. This methodology also works to capture, frame, and represent the essence of different layers of a location in a filmic narrative. In *Kanchenjungha* (1962),[15] different zones around the Darjeeling mall are delineated with certain recurring observational sonic details. When the characters are moving in or out from one zone to another, the locative sounds are changed accordingly. For instance, the central meeting place of the characters in the middle of the town has a complex soundscape with layers of multiple man-made sounds such as the murmurs of speech, footsteps, and a cycle's bell. While the characters are moving toward the territory of the town, up or down the hill, some sounds stand out such as a bird call and the sounds of crickets while Bannerjee and Manisha take their afternoon stroll along the hills. As they move from the mall to the central part of the town, the sound of a barking dog fades in, replacing the distant bird song and cricket sounds. These situated sound motifs are heard across

14 Excerpts from the interview www.youtube.com/watch?v=Aql4EREI1Ak (accessed 10 May 2023).

15 Link to the film: www.youtube.com/watch?v=8DOEibyLJhQ (accessed 5 July 2022).

multiple places, overlapping each other, and, in the process, they connect the places and provide an enveloping space for the narrative of temporal encounters and evolving relationships among the characters.

Furthermore, several sonic details accentuate the presence of settings such as the sound of horses riding past characters, the bells of a cattle procession passing by, and the prominent sound of the church bell, which suggest reconciliation between Anima, the eldest daughter and her husband Sankar. These sounds intersect with the perpetual meanderings of the characters, unifying the multiplicity of narratives in zonal separation while establishing a sense of the presence of Darjeeling's various locations with the gorgeous Himalayan peak of Kanchenjunga in the background. Taking another example, in *Mahanagar* (1963), the middle-class neighbourhood comes alive with a rich depth of sounds, ranging from children shouting on the street to temple bells, bicycles, rickshaws, and radios. All these sounds are off-screen, creating an expanded diegetic space around the northern Kolkata neighbourhood. They make the audience feel situated in the film space while being aware of an expanded universe swirling around, alive and vibrating. One can also critically listen to the sonic details of the household Gramophone in the third segment *Samapti* from *Teen Kanya* ('Three Daughters', Ray, 1961), based on three short stories by Rabindranath Tagore. The sound of the Gramophone operates as a portal to gauge the characters' interpersonal dynamics in their different contexts. In an earlier sequence, the young graduate Amulya returns home and plays a song on solitude ('Bosia Bijone') from a record while his mother asks him to get married. Later, the same Gramophone becomes a site of realizing nonverbal communication for the young bride Mrinmoyi, where she keeps the matrimonial jewelry. When she returns to the groom Amulya, she turns the Gramophone on without a record, and the repetitive sound of the empty Gramophone reverberates in the room where they reconcile and engage intimately.

In contrast, Ray's latter films appear relatively verbose, relying on the dialogue between characters to form the narrative. These films lack the realistic treatment of sound that invoked the sense of site or situated-ness demonstrated in his earlier films. Some Ray scholars (Sengupta, 2007 et al) have argued that his realist paradigm went through certain shifts. Evidence of this is prominent in his work collectively known as 'Calcutta Trilogy': *Pratidwandi* ('The Adversary', 1970), *Seemabaddha* ('Company Limited', 1971), *Jana Aranya* ('The Middleman', 1976). Alongside the conscious manipulation of imagery, Ray began to process sounds to serve impressionist purposes, creating situations that were emotionally bleak, nightmarish, or non-lyrical. Take for example, the flashback sequences of childhood in *Pratidwandi*

(1970):[16] the voice and the bird call in this sequence are processed with added reverb to evoke the inevitable passing of time and the association of memory that is distant and decaying in the present. In the last sequence of *Seemabadha* (1971),[17] protagonist Shyamalendu paces up the stairs and the sound of children playing on the small playground slowly disappears while his own footsteps become dominant until finally, at the door of his tenth-floor flat, we hear only the sonically filtered sound of his breathing. However, there are sequences where locative sounds take on a crucial role, suggesting an urban mood of alienation and miscommunication. Take for example, thick layers of sound from the city of Kolkata intruding into the silence as the characters meander on the terrace of a big building in *Pratidwandi* (1970). In one sequence in *Jana Aranya* (1976), Somnath and Sukumar are sitting near the Maidan area of Kolkata and the sound of traffic dominates their conversation, isolating the characters within their troubled friendship. A few of Ray's populist films from this period display hints of commercial inclinations – a mainstream storytelling strategy using stars instead of creating perceptually realistic atmosphere through sound. For example, in *Chiriyakhana* ('The Zoo', 1967), the multitude of traffic and city sounds of the opening credits succumb to narrative pressure. However, even these sonically weaker films from his latter period contrast sharply with the predominantly vococentric popular films produced at the same time in India.

An Audiographic Realism by Presence

As I have shown above, Ray's films generally highlight a distinct recognition for and a keen observation of the fictional site, narrated in the story-world by providing detailed locational information in the use of sounds. This narrative strategy operates within the apparent limitation of synchronized sound and the monaural framework, applied in Indian cinema's filmmaking processes and trajetories, from recording to post-production and to the film projection, of his day. Film theorist Noël Burch discusses similar limitations in American cinema:

> [In a monaural synchronized set-up, BC] microphone recording...would jumble them (sounds recorded from the location) all together; and

16 Link to the film: www.youtube.com/watch?v=-inaeJjndRg (accessed 5 July 2022).
17 Link to the film: www.youtube.com/watch?v=ghM7e2QIdaY (accessed 5 July 2022).

> the sounds emerging from the single source of the loudspeaker in the theatre would all be equally 'present', much as a camera reduces the three-dimensionality of real space to the two-dimensionality of screen space. (1985, p. 201)

In working within these limitations in Indian cinema and fully exploiting synchronized sound and monaural aesthetics while narrating his reality-based stories, Ray constantly strove to give due attention to the specifics and particularities of the site while constructing pro-filmic spaces in the diegesis. Site-specific sounds are the primary elements carrying a significance of the real people and places in this narrative strategy with realist sensibilities. His intention was to use the site as a character in many of his works. We learn more from the memoirs of Ray's wife, Bijoya (2011), and from Marie Seton's biography (2003) about Ray's fascination with the sites he chose to shoot at and his willingness to shoot on location for all his films. Besides his interest in shooting on location, he was also a great innovator with natural light available on site: along with his first cinematographer, Subrata Mitra, he pioneered 'bounce lighting' in *The Apu Trilogy* (1955–59) before Sven Nykvist claimed to invent the technique (Ray, 1978).[18] As discussed, Ray's frequent use of off-screen locative sounds suggests an awareness of the expanded presence of the site in his efforts to overcome the above-mentioned limitations of monaural synchronized sound recording and monophonic mixing. The practice was a departure from the classical conventions of the largely vococentric Indian cinema of the 1940s and 1950s.

Ray's strong belief in the capacity of actual sounds to carry the narrative lays the groundwork of what can loosely be called 'audiographic realism' in Indian cinema. By audiographic realism, I refer to the synchronized use of sound without significant sound manipulation that retains the materiality or the object-hood of the documented sound with perceptual fidelity, and a more realistic representation of the actual site and situation within a monaural screen-centric use of site-specific sound elements. This is analogous to photographic realism's determination not to affect the appearance of a photographic object (Kania, 2009, p. 240). Keeping a realist paradigm in cinema as an authorial choice, Ray used synchronized sound as a means to explore his cinematic ethos of audiographic realism, about which he states: 'the main contribution of sound was an enormous advance towards realism, and a consequent enrichment of the medium as an expression

18 See: seminar speech at American Film Institute, www.youtube.com/watch?v=TdFi1noweog&t=226s (accessed 1 March 2023).

of the ethos' (2011, p. 5). Here the definition of 'realism' refers back to the tradition of observational cinema, which represents reality by recording vision and sound that come 'from within the world of the film' (Kania, 2009, p. 244). Satyajit Ray, one of the most influential filmmakers from the Indian subcontinent, for whom the realist paradigm was an authorial choice, set a new benchmark in Indian cinema. Film scholar Anindya Sengupta comments on Ray's practice with naturalistic sound:

> Ray not only imbibed formal ideas from the dominant and emerging practices of realist cinema, but was also informed by other kinds of cinemas. He can certainly be credited for creating a model of narrative realism, which will be loosely followed by the succeeding art-cinema movements in India; ...the soundtrack in Ray's cinema became naturalistic (particularly in his use of 'voice', 'speech' and incidental 'noise'), reticent (in his famous use of silences or suspension of speech, variations in set themes and motifs of non-diegetic music) and auteuristic (i.e. it became eminently recognizable), compared to the more ornamental, generic and over-wrought instances of the mainstream melodramas. (2007, pp. 86–87)

As we have seen above, Ray's sound practice provides a detailed depiction of a location, and its environmental and social realities, to create a mode of situated listening in which the audience finds ample sonic information and aural associations to cognitively relate to the places and people as the very subjects of the narrative. In *Pather Panchali* (1955), a girl asks her brother to 'listen' as they stand at the side of a large field, and a layer of diegetic sounds follow. These sounds emanate from various objects within the landscape that are visible on screen: the electric pole, rain falling on the field, the passing train, the night cricket, various other insects, the forest, and so on. The sounds not only re-present these objects but also establish their very presence and their relational interplay with the characters who listen to the sounds, creating perceptual fidelity and depicting their situatedness in the cinematic universe in alignment with the lived experience. Film scholar Ravi Vasudevan speaks of this sense of presence and verisimilitude while discussing Ray in the context of contemporary filmmaking:

> The differences appear to emerge from evaluating the status of the narrative form through which the real would be articulated, through what means of representation, styles of acting, [and] aesthetic strategies the real would be invoked. Here, the popular compendium – studio shooting, melodramatic, externalized forms for the representation of character

> psychology, non- or intermittently continuous forms of cutting, diversionary story lines, performance sequences – was not acceptable within the emergent artistic canon, for they undermined plausibility and a desirable regime of verisimilitude. (2003)

Such an intent on providing spatial verisimilitude manifests itself not only in the elaborate use of sounds but also in the innovative techniques in recording and re-presentation of voices. In an interview, Ray's sound mixer Jyoti Chatterjee speaks about Ray's use of 'straight or direct dubbing' to avoid certain camera and production noises;[19] the method is used to capture the actor's voice immediately after the shot is taken in contrast to the standard loop dubbing used in Indian cinema from the 1960s onwards. 'Direct or straight dubbing' is performed in the same place and under similar circumstances to retain the spatial authenticity of the site as well as the site-situated performance of the actors, something that couldn't be recreated in the same way in the studio in front of the looped reference of visual images. We find evidence of such innovative dubbing techniques in *Nayak* (1966) and *Aranyer Din Ratri* ('Days and Nights in the Forest', 1970). This practice of random dubbing on location (to retain the on-site details and ambience when direct synchronized recording is confronted by logistic difficulties) shows Ray's commitment to creating a spatial presence in the sonic diegesis, expanding the possibilities of creative sound production in Indian cinema.

Ray's Sonic Legacy

Speaking of the legacy Ray left Indian cinema, Shyam Benegal states:

> Cinema was of a particular kind before Ray came. After Ray, it was something else in terms of looking at cinema itself. I'm not talking about popular cinema – that is different. But, when sound came, that was the first dramatic change that took place in Indian cinema. The second dramatic change in Indian cinema was the coming of Satyajit Ray. (Chattopadhyay, 2021b, p. 30)

It is no surprise then that in contemporary Indian cinema of the digital era, Ray's sonic sensibilities find a sort of revival. Many of today's films incorporate detailed layers of synchronized sound recordings with digital

19 See: interview with Jyoti Chatterjee in *Between the Headphones* (Chattopadhyay, 2021b).

multi-track capabilities and multi-channel surround mixing. This mode of production illustrates an emergent interest in representing real locations and spatially faithful auditory settings. Consequently, these sonic layers suggest the presence of actual cinematic sites in the narration, creating believable film spaces that use the intricate sound elements in a more spatially realistic manner. Ray's creative use of sound intended to delineate such a spatial practice and did so in the way he used the limited technical measures of monophony available to him. The detailed use of sounds in contemporary Indian films hints at a rediscovery of cinema's realistic origins dating back to Ray's film oeuvre. Therefore, in tracing Ray's position, it becomes possible to use his sound practices as a reference marker through which developments in film sound can be investigated. It is important to revisit Ray's works with awareness about his cinematic sound production ethos – particularly his use of spatial sound elements – in order to begin relocating his works within the contemporary milieu of Indian cinema, which celebrates complete digitalization.

References

Allen, R. (2009). "Pather Panchali". In Lalitha Gopalan (ed.), *The Cinema of India* (24 frames series, pp. 86–95). London: Wallflower Press.

Birtwistle, A. (2010). *Cinesonica: Sounding Film and Video.* Manchester: Manchester University Press.

Burch, N. (1985). "On the Structural Use of Sound". In E. Weis and J. Belton (eds) *Film Sound: Theory and Practice* (pp. 200–209). New York: Columbia University Press.

Chattopadhyay, B. (2015). "The Auditory Spectacle: Designing Sound for the 'Dubbing eEa' of Indian Cinema". *The New Soundtrack* 5(1), 55–68. Edinburgh: Edinburgh University Press.

Chattopadhyay, B. (2021b). *Between the Headphones: Listening to the Practitioner.* Newcastle upon Tyne: Cambridge Scholars Publishing.

Chion, M. (1994). *Audio-vision: Sound on Screen.* Translated and edited by C. Gorbman. New York: Columbia University Press.

Cooper, D. (2000). *The Cinema of Satyajit Ray: Between Tradition and Modernity.* New York: Cambridge University Press.

Ganguly, S. (2001). *Satyajit Ray: In Search of the Modern.* New Delhi: Indialog.

Gopalan, L. (2002). *Cinema of Interruptions: Action Genres in Contemporary Indian Cinema.* London: British Film Institute.

Lastra, J. (2000). *Sound Technology and the American Cinema: Perception, Representation, Modernity.* New York: Columbia University Press.

LaBelle, B. & Martinho, C. (eds) (2011). *Site of Sound: Of Architecture and the Ear: Vol. 2*. Berlin: Errant Bodies Press.

Kania, A. (2009) "Realism". In P. Livingston and C. Plantinga (eds), *The Routledge Companion to Philosophy and Film* (pp. 237–247). London and New York: Routledge.

Ray, B. (2011). *Amader Katha*. Calcutta: Ananda Publishers.

Ray, S. (2011). *Satyajit Ray on Cinema*, edited by Sandip Ray. New York: Columbia University Press.

Sengupta, A. (2007). "Seeing Through the Sound: Certain Tendencies of the Soundtrack in Satyajit Ray's Films of the 1970s". *Journal of the Moving Image* 6, 86–101.

Seton, M. (2003). *Portrait of a Director: Satyajit Ray*. New Delhi: Penguin Books India.

Sonnenschein, D. (2001). *Sound Design: The Expressive Power of Music, Voice, and Sound Effects in Cinema*. Studio City, CA: Michael Wiese Productions.

Vasudevan, R. (2001). "Nationhood, Authenticity and Realism in Indian Cinema: The Double Take of Modernism in the Work of Satyajit Ray". *Journal of the Moving Image* 2, 52–76.

6. Popular Films from the Dubbing Era

Abstract: In the sixth chapter, I discuss the filmmaking period from roughly the 1960s to the late 1990s in India known as the 'dubbing era'. Many representative popular films from this era are examined to investigate how the asynchronous practices of sound in this period incorporated a technologically informed approach, using analogue sound processing with expressionistic and melodramatic overtones. Following critical analysis of sound from several films and drawing supporting evidence from the conversations with veteran practitioners, the chapter demonstrates how magnetic recording and dubbing rendered the sound of this period as something spectacular, with extravagant song-and-dance routines in foreign locations packed with studio-manipulated, synthetic effects. Add to this a deliberate lack of site-specific sounds, and this practice triggered a cinematic experience of emotive tension and affective stimulation.

Keywords: dubbing, ADR, Indian cinema, playback system, song-and-dance routines

Loop Dubbing

From the late 1940s onwards to the digital era, sound in Indian cinema was rarely recorded on location. Dubbing and re-recording were deployed instead as the dominant modes of sound practice in mainstream Indian films; the film soundtrack was mostly created (or re-created) in the studio. Actors would recite and re-record their lines as their images appeared on the studio screen in a process known as 'loop dubbing', via 'looping',[1] or ADR (Automatic Dialogue Replacement). Background music, along with lavish song and dance sequences at regular intervals followed this process. Various sound effects, or Foley, were entirely made in a post-synchronization process

1 See: interview with Anup Mukherjee in *Between the Headphones: Listening to the Practitioner* (Chattopadhyay, 2021b).

Chattopadhyay, B., *Sound in Indian Film and Audiovisual Media: History, Practices, and Production*. Taylor & Francis Group, 2024
DOI 10.5117/9789463724739_CH06

within the close confines of a studio and added later as the secondary or tertiary layers of the sound organization.

The use of ADR in Indian cinema became a regular practice from the 1950s onwards with the arrival of the Arriflex 2C and Arriflex 3 cameras, which required the soundproof cover called 'blimp'. This distracting camera noise always entered the pilot track recorded on location and it required all the sound components, including character dialogues and synchronized sound effects, to be re-created in the studio. Eventually, this practice of loop dubbing became the standard in Indian films. As dubbing emerged alongside the standardization of analogue magnetic recording and mixing, it was facilitated by multi-track re-recording in the studio. The following phase of sound production in Indian cinema was shaped towards what is known as the 'dubbing era' (roughly between the 1950s and 1990s). This was a long stretch of time that illustrated a growing interest in the controlled deployment of a few sound elements as design material in films, keeping the primacy of the voice along with a prominent usage of background music, song-and-dance sequences, and processed sound effects. However, in this hierarchy of sound organization, there was a substantial lack of actual locational sounds. This practice was a result of the standardized methods of studio-centric film sound production with a particular aesthetics of post-synchronization using dubbing and Foley as methodologies.

Outside of the vocabulary of filmmaking, 'dubbing' generally refers to the replacement of the actor's voices by different performers or 'voice actors' speaking another language to make it more accessible to audiences abroad. In film terminology, this process is sometimes called 'revoicing'. The coming of sound divided Indian cinema into different linguistic territories and regions. But through dubbing, there was an effort to make regional cinema available across these territories. For example, Hindi-language films were dubbed into Bhojpuri. More specifically, loop dubbing, dubbing, or ADR are used in Indian cinema as a technological innovation to replace the voice recorded on location, acting as the pilot or guide track for the same actor's voice inside the controlled environment of a studio, in which incidental and camera sounds can be avoided. This is the aspect of voice replacement that I am focusing on in this chapter, as I am interested in uncovering the creative use of sound, especially actual location sounds.

In *The Auditory Setting* (2021a), I have discussed at length how the practice of dubbing impacted cinema in the production of a suspension of disbelief while listening to the film. There, I mention Jorge Luis Borges's take on dubbing as a crucial perspective, in which he refers to dubbing as 'a perverse artifice' contributing to 'ingenious audio-visual deformation' (1945, p. 262).

I have shown how his indignation with respect to dubbing is based on the voice replacement practices used by foreign exhibitors, whereby American films were converted from English to other languages. In India the replacement was also spatial and temporal, rather than just linguistic, since most of the time, the same actors would eventually dub their own in-sync voices (often in the same language), only to replace their directly recorded voice on the pilot or guide track during sited performance with post-synchronized voices recordings made inside sterile studios. The question, then, was not only what film scholar Nataša Ďurovičová terms as 'voice-body duplication in post-synchronicity' (2003)[2] but it was also a more asynchronous methodology to achieve cleanliness in voice recording at the expense of spatial and temporal credibility, having little or no audible information about the cinematic site narrated in the story. It was an odd experience listening to dubbed Indian films because the actors were divided in space and time while performing their lines of speech twice – once on location and again inside the studio. This process of voice replacement 'privileges one aspect of speech, namely comprehension, at the expense of all of its other aspects' (Ďurovičová, 2003). To compensate for this loss of actuality, which Borges calls 'audio-visual deformation' via a 'perverse artifice', voice, music, and sound effects were mixed loudly in an overly expressionistic style using analogue audio processing techniques and methods. Also, the frequent interruptions of song-and-dance sequences (Gopalan, 2002) were used to temporarily suspend the flow of narration for an indulgent escape. In *The Auditory Setting* (2021a), I discuss in detail how this aesthetic strategy maintained a primarily non-diegetic film space where actual locational sounds wouldn't fit in.

Commercial Indian films of this time were indeed known for their poor quality of sound design embedded with a lack of creative practice. As such, there was an apparent deficiency in the industry personnel's sonic sensibility to create an aurally perceivable film space and convincing cinematic spatiality, perhaps symptomatic of the 'peculiar inability of Indian cinema to produce a persuasive relationship with live location sound, the only proper sound resource actually available to the cinema' (Rajadhyaksha, 2007, p. 14). On the other hand, such uneven uses of sound may resonate with the traditional practices such as oral storytelling, and a thrust on narrative quasi-diegesis rather than a mimetic approach to authenticity and realism. Tejaswini Ganti hints at why the dominance of dubbing in Indian cinema can be attributed to India's traditional 'oral style of working'

2 See: https://epa.oszk.hu/00300/00375/00001/durovicova.htm.

(2012, p. 156). It is crucial to study these peculiar trends in sound practice and sonic methodologies manifesting in a lack of actual spatial sounds, the dearth of the presence of the site, and a disinterest in the convincing establishment of the pro-filmic space and the aural situated-ness in diegesis as the dominant narrative strategy.

Dubbed Voice and Post-Sync

If we critically listen to some of the popular films from the dubbing era's peak period, namely in the 1970s and 1980s, we notice the slippages and errors of post-sync dubbed voices. Particularly, in action-driven narrative films like *Zanjeer* ('Shackles', Mehra, 1973), *Amar Akbar Anthony* (Desai, 1977), *Deewaar* ('The Wall', Chopra, 1975), *Lahu Ke Do Rang* ('Blood Has Two Colours', Bhatt, 1979), *Yaadon Ki Baaraat* ('Procession of Memories', Hussain, 1973), and *Parvarish* ('Upbringing', Desai, 1977). These films prioritize a loud and dramatic use of sound (e.g., inordinate musical scores and modulated sound effects) at the expense of a well-balanced, mindful, nuanced, and ecological organization of sound. In this framework, the voice is considered as the primary sound component to carry the lavish storyline; hence, it is dubbed in the studio and treated during post-sync for clarity, compromising the auditory setting of the voice in the process. For example, in *Zanjeer* (1973), the protagonist Vijay, in his first scene as a police officer, speaks with his senior police commissioner, Singh. In their conversation, set in the indoor police office, Vijay's voice is recorded directly as in straight dubbing, and Singh's voice is loop-dubbed. Due to this discrepancy, Vijay's voice carries a spatial ambience while Singh's is clinically clean. They carry audibly different spaces in their voice while sitting on opposite sides of a table in the same room. This spatial difference doesn't add to the complexity of their interaction, so it is hard to believe that it is intentional and a part of the script. This was most likely a slippage like many other similar inconsistencies and 'mistakes' that a sensitive cinema goer would be able to easily locate in many films from this era.

The narrative strategy of practising sound in Indian films of the dubbing era needs to be understood through the lens of specific aesthetics of analogue (magnetic) sound technologies and aesthetics (Lastra, 2000; Sonnenschein, 2001) applied in cinema. This draws from studies in (film) sound production (Buhler, Neumeyer, and Deemer, 2010; Sergi, 2004) and its influence on stylistic features in cinematic sound (Biancorosso, 2009; Lastra, 2000), and can be applied to Indian film sound and screen music production

(Pemmaraju, 2013; Vijayakar, 2009). The representation of sound in overly expressionistic, spectacular, and melodramatic overtones that distanced cinematic sound's actuality from its pro-filmic spaces and sites – and in the process created emotive and affective responses in the mise-en-sonore or auditory setting – is understood using theories of emotion and affect in films (Tan, 2011; Plantinga, 2009). Additionally, the typical lack in the creative use of sound at this time is observed in light of narrative strategies of 'diegesis' (Percheron, 1980; Burch, 1982) and by considering the notions of 'presence' drawn from the studies of sound perception and cognition (Lombard and Ditton, 2006; Skalski and Whitbred, 2010; Grimshaw, 2011). Although the above body of scholarly literature was developed by primarily keeping examples of American and/or European films in mind, these references create an epistemological grounding of film sound studies where the examples from diverse modes of Indian cinema can also be contextualized, historicized, and theorized. In this book, these perspectives work as a heuristic for Indian cinema.

Recorded Tape as Raw Material

During the 1960s, magnetic recording began to be extensively used in Indian cinema. With the arrival of magnetic tapes, sound recording became easier to do and was of a relatively better quality. The magnetic tape-based recording machines became portable, making it possible to copy, store, and erase tracks of recorded sounds whenever required, even at the expense of asynchronous modes,[3] since magnetic tapes weren't coupled with the camera. Unlike the previous direct optical recording system, this situation created scope for mixing and re-recording beyond the location. James Buhler, David Neumeyer, and Rob Deemer comment on the advantages of analogue magnetic sound recording in films: 'The introduction of magnetic tape likewise allowed an efficient and relatively inexpensive way to record and mix sound into a number of channels. This also provided an efficient way to provide a variety of mixes' (2010, p. 338). In Indian cinema, as in Hollywood, the advent of magnetic tape made a better quality of recording possible, with a higher dynamic range than that of the previous practice of direct optical recording.[4] Indian film writer Gautam Pemmaraju observes that 'magnetic tape revolutionized sound recording and production, and

3 Separately recording various film sound components almost entirely inside a studio.

4 See: interview with Anup Mukherjee in *Between the Headphones* (Chattopadhyay, 2021b).

it had a significant impact on films. Sound captured analogously on tape with magnetic emulsion on it was of far better quality and dynamic range than that on optical film' (2013, p. 80). The 35mm optical filmstrip had a dynamic range of 78 dB, which was what the optical direct sound recording could get as the 'headroom'[5] of recorded sound, limiting the signal-to-noise ratio and dynamic range. Within this narrower dynamic range, the vococentric recording naturally prioritized the most legible recording of the voice possible, delimiting other sound components such as incidental and location-specific sounds for the film soundtrack, putting an emphasis on the voice as the primary narrative vehicle. In the magnetic era, the dynamic range of (magnetic) sound recording was around 98 dB (depending on the material). This wider headspace allowed for an inclusive capacity for recording, layering, designing, and mixing sounds. This also paved the way for dubbing to take centre stage.

Following the emergence of magnetic recording, in-house sound studios became popular for doing post-production instead of direct recording on location. As a result, film sound became increasingly distant from the real site narrated on screen. Gradually, an analogue studio-centric technique of film sound design emerged as the dominant mode of practice. Dubbing and Foley followed technological advancements, including the introduction of the Nagra portable recorder[6] (utilizing quarter-inch magnetic tape) and the MagnaTech Rock-and-Roll mixing console.[7] The studio system invited more investment in sound post-production. Tools and techniques like looping, multi-track mixing, and track-laying opened up possibilities of parallel resourcing of sound reconstruction, making the dependence on direct location recording obsolete. Stock sounds were shared or became commercially available as a bank of 'sound objects' from which one could pick up raw materials for ambience and sync sound effects, although, in most cases, site-specific sounds such as ambience were a minor concern in sound organization. With the rise of the new technologies of analogue magnetic recording, the sound objects recorded on magnetic tape were no longer referred to as a site-specific sound source (Demers, 2009) as

5 As mentioned earlier, 'headroom' means the amount of loudness that exceeds a designated reference level a sound signal can handle before it distorts or clips.

6 Mostly battery-operated and 'phantom' (48v) powered, portable, professional audio tape recorders produced by Kudelski SA, based in Switzerland. The Nagra II model has been the most popular of the series for use in the Indian film industry.

7 Sound designer Dipankar Chaki speaks of the early years of the MagnaTech Rock-and-Roll mixing console in the Bengali film industry and how that improved the sound quality. See: interview in *Between the Headphones* (Chattopadhyay, 2021b).

discussed earlier in the second chapter. I showed in this chapter that in Indian cinema, the practice of recording sound on tape enhanced sound's materiality so as to make it possible to decouple sound from the site as a raw material. At this point, one could bring these raw materials into the studio to follow processes of analogue post-synchronization, re-coupling it with the image asynchronously. Following this method, films were increasingly shot on a pre-designed set inside studios instead of in real locations, and film sound production was becoming a mere product of a dialogue-background, score-sync effect production chain.

Thus, a practice was carried out by industry-dependent and technologically informed sound technicians to design a soundtrack for a film out of asynchronous sound sources, using pre-recorded sound materials from sound banks. In many films the same sounds taken from the sound bank would be used in loops, such as birdcall, traffic noises, crowd murmurings, and so on. Any careful listener could locate these primitive loops and the repetitions of a particular stock sound in many different films. The same bird would sing in a loop or the same crowd would feature in the outdoor scenes from several films with different narratives. In most cases, these sound technicians didn't give much attention to the creative and nuanced use of sound, which would have otherwise demanded closer attention to the location or the source of sound in its recording and design stages. Song-and-dance sequences and loud background music were frequently used merely to mask these shortcomings, as veteran sound mixer Anup Dev informs us.[8] Many sound practitioners admitted that the use of a loud mix in the background score limited the possibility of including any substantial layers of locative sounds in the soundtrack. The lack of ambient sounds meant a lack of information about the pro-filmic space in the narration (Burch, 1982). This trend of analogue studio-centric sound production in the dubbing era thus approached a relative abstraction of the film's diegetic universe.

Sound without Perspectives

As discussed, in order to avoid camera noise emanating from the popular and affordable Arriflex 3 cameras along with other incidental background noises from the location of shooting, loop dubbing was introduced in Indian cinema and became a norm thereafter. Loop dubbing used re-recording techniques inside insulated sound studios. As previously mentioned, this

8 See personal interview with Dev in *Between the Headphones* (2021b).

process became known as ADR, and required actors to perform twice: once on location in front of the camera, and again in the studio in front of a microphone. Such sound practices in the dubbing era with emphasis on post-synchronization effectively destroyed 'perspective' in cinema (Pemmaraju, 2013), in the sense that the physical reality of the sites with their spatial details and natural auditory situations from the location could not be re-created inside the sound studio.

> A critical landmark in the history of cinema is the separation of the production of the sound from the image. In India, this led to the 'playback' and later, the 'dubbing' era. ...The production of the dialogue also came to be separated from the filming on a set or a location. This was driven by a technological imperative: the noise made by the moving parts of the camera. (Pemmaraju 2013, p. 80)

If we follow the dialogue delivery in some of the representative popular films of the 1970s, such as *Dharmatma* (Khan, 1975), *Johny Mera Naam* ('Johny My Name', Anand, 1970), *Deewaar* (Chopra, 1975) and *Lahu Ke Do Rang* (Bhatt, 1979), we can hear the actors' voices processed with studio reverb even in natural outdoor settings. In a particular sequence of *Dharmatma*, the tribal woman Reshma and the main protagonist Ranbir, who has fallen in love with her, speak in an open clearing in the middle of a forest (according to the storyline, the site is in the distant land of Afghanistan). We hear the actors' voices which have been processed clean, polished, and made crisp using a sound compressor to accentuate intelligibility. We don't hear the ambience of the forest, nor do we perceive the auditory perspectives and the depth of sonic field of their respective positions in this sonic field as they stand facing each other on either side of a tree. Later, when the villain Jankura arrives, envious of their amorous proximity, he shouts in anger, and his voice emerges, treated with the acoustic processing of a reverberant room rather than the openness of a forest. Similarly, in an earlier sequence from *Deewaar*, someone makes a speech in an exterior courtyard in front of a number of factory workers, but his voice appears cleanly recorded by a method of close micing with applied interior studio reverb, without the slightest sign of the auditory perspective his position in an open field should create. There are numerous such examples where the voice lacks a well-defined auditory perspective and a convincing depth of field based on the relative positions of the characters in the pro-filmic spaces.

James Lastra has shown that, in American cinema, sound dubbing and its use of the 'close frontal miking of actors, which minimizes reflected and indirect sound, became the norm for dialogue' (2000, p. 142). Lastra acknowledges the lack of acoustic perspective this specific practice creates: 'Despite changes in shot scale, the sound recordist has maintained "close-up" sounds in order to ensure intelligibility, thereby violating the presumed norms of "sound perspective"' (2000, p. 143). It is no surprise that these Hollywood norms would influence popular Indian cinema too, exporting similar production aesthetics into the Indian context with disastrous results.

Modulated Effects

The car-chase sequence in *Deewaar* (1975) and the fight sequence in *Dharmatma* (1975) exemplify the use of sound effects in an overly enhanced and dramatically modulated style, rendered with hyper-real textures. When the car of the protagonist, Bijay, reaches the end of its fateful journey at the top of the temple stairs, the tyres make a loud and un-naturalistic sound of screeching that wouldn't occur in a real situation, since the car stopped racing up the concrete stairs after leaving the road. The fighting actors' punches and the gunshots coming from their pistols are processed with extra reverb, time-stretching,[9] and compression[10] to render the sounds histrionic. They do not sound 'real' at all, even though the film deals with the hard fought social realities of India. What we hear is a hyper-real environment of suspended disbelief and emotive tension that doesn't realistically relate the sound to the locative source depicted in the cinematic universe. Narration in these films 'creates a specific emotional tension' (Tan, 2011, p.35) via studio manipulation, and the abstraction and modulation of recorded sound materials as processed audio effects, producing a suspended and 'unsitely' reality.

It would be interesting to see how parallel practices were operating in music production at this time and see if there was any link between film sound production and Indian music industries. Rajiv Vijayakar, in *The History of Indian Film Music* (2009), has pointed out how such hyper-processed sonic modulations were commonplace in the production of soundtracks in Indian cinema much like they were in music production:

9 See: interviews with Anup Dev and Hitendra Ghosh in *Between the Headphones* (Chattopadhyay, 2021b).

10 'Time-stretching' is used to make sound effects appear dramatic and temporally intriguing. And 'compression' is employed to smooth out sound's rough edges.

> The foremost changes that came in this decade were probably the advent of consciousness in sound and the beginning of change in the old order, and both these factors were also interconnected. ...As a kind of compensation, echo or reverb was added to...film soundtracks. (2009, p.53)

Thus, in the dubbing era of Indian cinema, the processing of sound effects (e.g., echo reverb, time-stretching, synthesizing, and EQ) also used in film music production was the result of consciously working with sound to control and compensate for the lack of realistic sonic representations. This can be understood as the films' asynchronous tendencies; separating sound from its real site-specific sources enhanced its dramatic and spectacular qualities. As discussed earlier, this perspective on consciously processing sounds seems comparable to what Emily Thompson has articulated in the context of American cinema as 'a fundamental compulsion to control the behaviour of sound' (2004, p. 2).

It is no surprise that the gradual standardization of sound production during the dubbing era of Indian cinema led to the creation of popular and syncretic formulations like 'Bollywood', which otherwise have little significance other than a normative popularization of the interrupting song-and-dance sequences, and instigation of dramatization in sound design to enhance sound's emotional and affective qualities, playing on the fringes of audiences' emotional reception. The intention has been to create popular mass appeal by exploiting melodramatic overtones through over-processing of the character's voices and modulating, for example, a villain's bodily sound effects. The specific sonic representation of a Bollywood villain's character is constructed using vocal manipulation tools like compressors, and by extending the reverb of footsteps and other bodily gestures, thus affecting visceral responses in the audience by affective mimicry – affecting the spectator as 'auditory entrainment' (Plantinga, 2009, p. 94). This production aesthetic became an accepted style to develop the scope of the cinematic spectacle in the following stages of sound practices, transitioning from monaural to stereophonic mixing.

The Great Sound Escape

If we study the films made at the peak of the dubbing era during the 1970s and 1980s, common threads would include a relative absence of spatial sound components in their sound organizations, and a weaker audio-audio relationship in terms of balance, ecology, and volume. In *Disco Dancer* (1982), the boy protagonist is caught with a stolen guitar in a park, but we do not hear any ambient sound that should be present in a wide-open public space. In *Maine Pyar Kiya* ('I've Loved', 1989), Karan's daughter, Suman, comes from the village to stay in a big city, but, apart from the sound transition of a train, there is no ambient sound layer testifying to the presence of a noisy, urban Indian environment. Sound mixing specialist Promod Thomas sheds light on the prescriptive practices for creating stereo sound mixes in Indian films during the dubbing era, avidly followed by sound practitioners. Convention dictated that the voice occupied the centre of the screen while sound effects and background music were placed in the side speakers of the stereo soundtrack. In this standardized scheme of sound organization, there was no place for ambient sound layers.[11]

Béla Balázs stated that film sound's role is 'to reveal for us our acoustic environment, the acoustic landscape in which we live' (1985, p. 116). Likewise, the relative absence of locative sounds denies the very revelation of acoustic environments and settings in which these stories take place. In *The Auditory Setting* (2021a), I have shown, with many examples, how locative environmental sound provides the depth of a shot by establishing the association between the viewer and the site, reinforcing 'the impression of reality' (Percherron, 1980, p. 17) in the processes of diegesis. This is achieved by providing a testimony of the site in the perception of direction and localization, as well as acoutsic perspectives; the spectator can relate to the sonic environment or the auditory situation of the pro-filmic space. Sound and music scholar Mark Grimshaw shows that adding locative sounds in the narration 'can create an immediate experience of presence and reality' (2011, p. 32). These sound elements, if used, supply sonic layers of realistic depth of field and perspectives replacing the one-dimensional, flat surface of the soundtracks associated with Indian cinema of the dubbing era. It was no surprise that voice and processed sound effects dominated practices during the dubbing era and merely served the visual authority of the film narrative instead of creating a multi-modal sensorial experience. In other words, the method of dubbing offered what film scholar Giorgio Biancorosso articulates as:

> [T]he illusion of a sumptuous, perceptually vivid impression of a causal relation which is known to be purely imaginary – one that is forced down

11 Sound mixer Promod Thomas stresses that in mono as well as in stereo track-laying, adequate space is not there for a fuller inclusion of ambient materials. See: interview in *Between the Headphones* (Chattopadhyay, 2021b).

> our throats, in fact. As such, dubbing stands in spectacular contrast to our everyday experience of a great many causal relations that, though known to be scientifically true, cannot be grasped through our senses. (2009, p. 267)

In contrast, according to theories of presence (Lombard and Ditton, 2006; Skalski and Whitbred, 2010), site-specific sounds can affect spatial presence by instigating the sense of 'being there' in the embodied experience of cinematic environments. This is achieved by presenting vivid and elaborate information about a site that places the audience within the film space and contributes to the narrative process of diegesis with a sensation of reality (Skalski and Whitbred, 2010). The absence of such creative layers of sound, therefore, renders the cinematic site imaginary – audiences cannot bodily or cognitively relate to the site they encounter in the story. The dubbing era of sound production in Indian films provides such an experience. Keeping locative sounds at bay, Indian films of this period predominantly create a remote and imaginary film space through a spectacular experience of momentary escape from the pro-filmic spaces, from aural situatedness, social realities, and the burden of locative presence.

Through a sonic phenomenological survey of a few significant examples of Indian films of this long period (1950s–1990s) and by taking evidence from conversations with Indian film sound practitioners, I have traced specific factors that informed sound practices in the dubbing era of Indian cinema. Such a sound practice led to a loud use of music and songs, dubbed voices and the minimal use of actual sounds to create an auditory setting that predominantly reproduced a sense of hyper-realism and remoteness through a site-unspecific sound narration and a relative absence of spatial sensibilities.

References

Balázs, B. (1985). "Theory of the Film: Sound". In Weis, E. & Belton, J. (eds), *Film Sound: Theory and Practice* (pp. 116–125). New York: Columbia University Press.

Biancorosso, G. (2009). "Sound". In Plantinga, C.R and Livingston, P. (eds), *The Routledge Companion to Philosophy and Film* (pp. 260–267). London: Routledge.

Borges, J. L. (1999). *The Total Library: Non-fiction 1922–1986*. London: Penguin Publishers.

Buhler, J., Neumeyer, D., and Deemer, R. (2010). *Hearing the Movies: Music and Sound in Film History*. New York: Oxford University Press.

Burch, N. (1982). "Narrative/Diegesis – Thresholds, limits". *Screen* 23(2), 16–33.

Chattopadhyay, B. (2021a). *The Auditory Setting: Environmental Sounds in Film and Media Arts*. Edinburgh: Edinburgh University Press.

Chattopadhyay, B. (2021b). *Between the Headphones: Listening to the Practitioner* (2021b). Newcastle upon Tyne: Cambridge Scholars Publishing.

Demers, J. (2009). "Field Recording, Sound Art and Objecthood". *Organised Sound* 14(1), 39–45.

Ďurovičová, N. (2003). "Local Ghosts: Dubbing Bodies in Early Sound Cinema". *Moveast* 9. https://epa.oszk.hu/00300/00375/00001/durovicova.htm (accessed 10 September 2023).

Ganti, T. (2012). *Producing Bollywood: Inside the Contemporary Hindi Film Industry*. Durham, North Carolina, United States: Duke University Press.

Gopalan, L. (2002). *Cinema of Interruptions: Action Genres in Contemporary Indian Cinema*. London: British Film Institute.

Grimshaw, M. (2011). *Game Sound Technology and Player Interaction: Concepts and Developments*. Hershey and New York: Information Science Reference.

Lastra, J. (2000). *Sound Technology and the American Cinema: Perception, Representation, Modernity*. New York: Columbia University Press.

Lombard, M. and Ditton, T. (2006). "At the Heart of It All: The Concept of Presence". *Journal of Computer-Mediated Communication* 3(2). https://academic.oup.com/jcmc/article/3/2/JCMC321/4080403 (accessed 10 September 2023).

Pemmaraju, G. (2013). "Soundbaazi: The Sound of More than Music". *ArtConnect: The IFA Magazine* 7(1), 62–82.

Percheron, D. (1980). "Sound in Cinema and its Relationship to Image and Diegesis". In Altman, R. (ed.), *Cinema/Sound* (pp. 16–23). New Haven, CT: Yale French Studies.

Plantinga, C. (2009). "Emotion and affect". In Plantinga, C. R. and Livingston, P. (eds) *The Routledge Companion to Philosophy and Film* (pp. 356–365). London: Routledge.

Rajadhyaksha, A. (2007). "An Aesthetic for Film Sound in India?". *Journal of the Moving Image* 6. https://www.jmionline.org/article/an_aesthetic_for_film_sound_in_india (accessed 10 September 2023).

Sergi, G. (2004). *The Dolby Era: Film Sound in Contemporary Hollywood*. Manchester: Manchester University Press.

Skalski, P. & Whitbred, R. (2010). "Image versus Sound: A Comparison of Formal Feature Effects on Presence and Video Game Enjoyment". *PsychNology Journal* 8(1), 67–84.

Sonnenschein, D. (2001). *Sound Design: The Expressive Power of Music, Voice, and Sound Effects in Cinema*. Michigan: Michael Wiese Productions.

Tan, E. S. (2011). *Emotion and the Structure of Narrative Film: Film as an Emotion Machine*. New York and London: Routledge.

Thompson, E. (2004). *The Soundscape of Modernity: Architectural Acoustics and the Culture of Listening in America, 1900–1933*. London: The MIT Press.

Vijayakar, R. (2009). *The History of Indian Film Music: A Showcase of the Very Best in Hindi Cinema*. New Delhi: Times Group Books.

7. Parallel Sounds, Radical Listening – Part I

Abstract: In the seventh chapter, I discuss the sound practices and sonic aesthesis emerging from the various strands of 'Indian New Wave'. Also known as Parallel Cinema, or arthouse cinema, this stream of films produced predominantly from the 1960s to the early 1980s, largely supported by the state, identifying with experimentation, certain aesthetic sensibilities, socio-political awareness, and innovative styles and techniques, was distinct from the commercial mainstream. Focusing on the works of pioneering auteurs such as Mrinal Sen, Ritwik Ghatak, Mani Kaul, Shyam Benegal, Adoor Gopalakrishnan, G. Aravindan, Kumar Shahani among others, this chapter shows how the use of sound from this era was a departure from that of the Golden Age. Sound began taking a central, creative role in film narrative.

Keywords: Indian New Wave, Parallel Cinema, arthouse cinema, alternative film movement

Sound after Ray

Indian cinema arguably touched modernity when the sound of a train gradually invaded the pastoral landscape in Satyajit Ray's *Pather Panchali* (1955). In this moment, the two protagonists, brother and sister Apu and Durga, lose their sense of orientation in a field on the edge of their village. The sound of the train, a new-fangled sonic intervention, destabilizes their sense of the everyday, bridging the previously uncrossed divide between a rural environment and the big city via a sound stage. The sound also transcends a formulaic period of Indian cinema, then known for its gargantuan output with arguably little substance. The film hints at the possibility of Indian cinema's future directions, inspiring generations of filmmakers towards an alternative and independent cinematic style, subverting the mainstream.

Chattopadhyay, B., *Sound in Indian Film and Audiovisual Media: History, Practices, and Production*. Taylor & Francis Group, 2024
DOI 10.5117/9789463724739_CH07

As shown in previous chapters, the 1950s and 1960s are considered the Golden Age of Indian cinema, when film auteurs such as Ray, Chetan Anand, and Guru Dutt emerged, putting Indian cinema on the international stage. At the same time, younger generations of filmmakers were initiating a new Indian independent cinema that assimilated wider influences, including European cinema, Indian folk, and vernacular cultures, drawing often from local myth and ritual practices.

The use of sound in these Indian New Wave films was a departure from their Golden Age predecessors. It began taking a central, creative role in the narrative. As we learn from interviews with Indian sound practitioners,[1] this was the time when sound production was evolving due to early technical advances. For example, audio effects such as reverb, EQ, compression, limiting, and echo were developed at this time. 'Looping', the technique that would quickly become standard industry practice to create dramatic effects, eventually followed.[2]

The so-called Indian New Wave has sometimes been termed Parallel Cinema, art house cinema, alternative film, etc. to reinforce the qualitative and aesthetic differences between these films and the commercial mainstream, both in Bombay and regional industries. These parallel streams of films can be understood as those produced between the late 1960s and early 1980s, largely supported by the state subsidy mechanism, identifying with experimentation, certain aesthetic sensibilities, socio-political awareness, and innovative styles and techniques.

Arguably, the forerunner of this movement was Satyajit Ray, along with his close contemporaries, namely Ritwik Ghatak and Mrinal Sen: the trinity from Bengal. However, Ghatak and Sen represent a more radical departure in terms of their use of sound. As discussed earlier, socially conscious and social realist films had previously been made, notably by Chetan Anand and Guru Dutt, Bimal Roy, K.A. Abbas, and V. Shantaram, but their works still belonged to the mainstream cinema of the time. *Pather Panchali* (1955) from Ray's *Apu Trilogy*, Ghatak's *Nagarik* ('The Citizen', 1952), his *Partition Trilogy* and *Ajantrik* ('The Unmechanical', 1958), and Sen's *Bhuvan Shome* (1969) shook the formulaic world of Indian mainstream cinema, introducing

1 As mentioned in the introduction, for this research I have interviewed prominent sound practitioners working in the Indian film industry to explore and understand their methods. Selected interviews from these long fieldworks are published as a parallel book: *Between the Headphones: Listening to the Practitioner* (Chattopadhyay, 2021b).

2 See: interview with sound designer Anup Mukherjee in *Between the Headphones* (Chattopadhyay, 2021b).

new styles and aesthetics in radically different ways, challenging viewers and changing the way Indian films were perceived locally and globally.

The next generation of filmmakers, who took this alternative, independent spirit of Indian cinema forward, include Mani Kaul, Shyam Benegal, Adoor Gopalakrishnan, and G. Aravindan. M.S. Sathyu, Kumar Shahani, Saeed Akhtar Mirza, Govind Nihalani, John Abraham, Shaji Karun, Girish Kasaravalli, Girish Karnad, Sai Paranjpe, Aparna Sen, and Ketan Mehta also contributed to this parallel film canon. Their films positioned the social issues and struggles of everyday life within a wider political context. Despite similarities in their socio-political stance, each are distinctive auteurs with individual artistic signatures.

Mrinal Sen and the Sound of Discontent

The period between the late 1960s and late 1970s constitutes the early phase of India's Parallel Cinema movement, possibly at its most creative and original. Sen set out to accomplish, both aesthetically and politically, an experimental style of filmmaking which freely used sound as a creative tool. *Bhuvan Shome* (1969) utilizes overly expressionistic and processed sound mixed in multiple layers to polemical, comical, and satirical effect. On Sen's processed sounds and image, film scholar Megan Carrigy states:

> The approach to storytelling in *Bhuvan Shome* offers up a dizzying array of interrelated non-naturalistic devices. Disjunctive, shock-producing storytelling techniques keep the audience in a state of anticipation. Techniques include merging voice-over narration and central character's internal dialogue, stark freeze frames, …rapid editing techniques such as jump cuts, abrupt changes in tone and tempo; and a complex soundtrack of railway sound effects. (2009, p. 142)

The flexibility of the newly available, tape-based, magnetic recording medium and its techniques were the conducive context for experimenting. In a documentary interview, Sen said (2012) that he exhaustively used the latest analogue technology available during the 1960s in India, incorporating studio techniques to their fullest during the making of *Bhuvan Shome* (1969).[3] With this film, the earlier trend of synchronized audiographic realism in Indian cinema took a serious turn toward non-naturalistic, processed, and

3 Aired on DR K (2012).

modulated sound, as previously trialed by Ghatak in his elaborate use of edited sound effects to create epic and melodramatic environments.

Bhuvan Shome (1969) begins with a close-up of fast-moving railway tracks. Responding to the engine's rhythmic tempo, a multitude of jarring musical chords are introduced utilizing the tabla, dholak, and other percussion instruments. Along with this unusual extra-diegetic soundtrack, a female voice starts singing the same basic seven notes at a lower register and pitch but maintaining the tempo, creating an expressionistic statement about narrative intervention grounded in reality.

In *Khandhar* ('The Ruins', Sen, 1984) sound effects add layers of parallel or auxiliary narrative streams. The film opens with the proximate sound of water splashing in a photo-developing tray. The sustained sound of splashing as the photographer tries to bring an image to life encourages the audience to have an indolent view of the emerging face of a woman, whose suggested melancholia will linger throughout the film. The sound of the photo being developed creates a subdued drama of anticipation, much of which is contributed by the sound. At the end of the film the image of the woman is seen mounted on the wall amid sounds of busy traffic in a city. As she stares from the photograph with her disenchanted eyes, the city responds with an inert heartlessness and forgetfulness.

Sen's more explicitly politically generative works such as *Interview* (1970), *Calcutta '71* (1972), and *Padatik* ('The Foot Soldier', 1973) chart India's wider historical, social, economic, and political context in malleable filmic form. The tumultuous times he tackled include: the Cultural Revolution unfolding in West Bengal throughout the 1960s; anti-establishment protests and street demonstrations by students who felt that neo-colonial systems needed disassembling; an influx of refugees from the 1971 Bangladesh genocide; and the ultra-left Naxalite Movement in the late 1960s that galvanized the political imagination of a generation of unemployed and disillusioned youth. Matching these contexts, Sen's sonic methodology involved shooting on location, in the busy streets of Calcutta, in a guerilla filmmaking style, tracing street demonstrations and capturing the public discontent of those times. In *Mahaprithibi* ('World Within, World Without', 1991), the film's profoundly disenchanted daughter hits an empty plate with a spoon, making a sharp, repetitive sound, until her father indignantly asks her to stop. Through such sonic interventions, Sen tended to reject a straightforwardly realist style of filmmaking that had become quite synonymous with the work of Ray and other Indian filmmakers of his time; rather he was looking for an evolving form of cinema in which self-reflexibility builds elliptical narratives capable of translating the anger and discontent of his time through celluloid sounds.

Ritwik Ghatak and the Shock of Post-Immersive Listening

Ghatak's first film, *Nagarik* (1952), though made before Ray's *Pather Panchali* (1955), was only released posthumously in 1977. Indeed, Ghatak struggled to find support for his films and to reach a wide audience. Though deeply powerful, compassionate, humanist, and revolutionary in spirit, Ghatak's journey through Indian cinema was a tragic one, much like the mythic quality of his cinematic oeuvre.

Experiments with sound and music are at the core of Ghatak's authorship. He frequently presents sudden, unexpected sound and moving image combinations to shock the audience. These quasi-diegetic juxtapositions operate beyond the 'illusion of redundancy' yet create the possibility for synchresis or 'the spontaneous and irresistible weld produced between a particular auditory phenomenon and visual phenomenon when they occur at the same time' (Chion, 1994, pp. 63–64). Ghatak's shock therapy challenges the audience's auditory comfort-zone and, through this rupture, helps them empathize with the film's protagonists and their situations. As an artist with firsthand, embodied, and situated experience of colonial atrocities and the violence of Indian modernity, Ghatak was obsessively focused on the profound trauma of the 1947 Partition of India, and the resultant mass migration of people to Bengal. And yet, his protagonists and situations, which follow these untoward political events in Indian history, don't merely document or retell people's plight. Ghatak was inclined to connect these events to meta-historical contexts of Indian customary life and ritual on a mythical and epic level. His films are a 'post-immersive' exercise in which audiences are forced to confront trauma through the temporal disjuncture of events by their critical and empathic faculties.

Drawing from my previous research on narrative immersion and taking a critical position on what I term 'post-immersion' (2020, 2021a), I underscore a disjunctive moment in a cinematic or medial experience where the alert subjectivity of an audience is encouraged to take form. Philosopher Vilém Flusser's notion of 'homeland' and 'homelessness' suggests that only when a person is removed from their place of origin do they become aware of their home ties, which reveal themselves as unconscious judgments. The idea of homelessness is useful here for thinking through post-immersion as a cinematic/sonic strategy, particularly in Ghatak's work. It is no wonder Ghatak was influenced by Brechtian Alienation Theory, which can be interlinked with Flusser's idea of homelessness. Cultural theorist Thomas Stubblefield notes:

> From Brecht's conception of epic theatre, Ghatak would borrow the dramatic technique of *Verfremdungseffekt* (translated a number of ways, most commonly as 'distanciation' or 'alienation effect'). In its original conception, distanciation referred to various formal techniques used in the theatre to discourage traditional identification with characters and onstage action in general in order that the audience might instead develop a critical and questioning relationship to what they were observing. In terms of the cinema, this technique often revolves around a 'foregrounding of the apparatus' or other meta-level cues that call attention to the constructedness of the image in order to similarly dissuade indiscriminate acceptance of the image as reality. (2006, p. 19)

Ghatak's sonic methodology of distancing or alienation manifests in the use of quasi-diegetic sounds within the narrative flow of events, thereby seeking the audience's participation in this critical distancing. It is only when the audience withdraws from the immersive experience and gathers their critical faculty that cathartic moments appear through this dramatic rupture. Take, for example, the sequence in *Subarnarekha* ('The Golden Thread', 1965) – part of Ghatak's *Partition Trilogy* along with *Meghe Dhaka Tara* ('The Cloud-Capped Star', 1960) and *Komal Gandhar* ('E-Flat', 1961) – in which the protagonist Iswar's broken family walks towards their new house by the river Subarnarekha for the first time. Iswar's little sister, Sita, asks the foreman, who came to pick them up from the station, about their new house. In the background we hear sounds of temple bells and conch shells, both associated with the ritualistic and mythical consecration of houses. In the same film, the news of the assassination of Gandhi reaches a refugee colony in Calcutta and someone in a very British accent shouts 'Hey Ram', which are considered Gandhi's last words. In another sequence the sonic rendition of 'Hey Ram' is used abruptly in the background while narrating the grown-up Sita's suicide in the presence of Iswar, who goes to her house unknowingly in search of sex, mistaking her for a prostitute. Personal tragedy thus merges with political tragedy, shaking up and disrupting the audience's narrative expectations.

A highly cited use of rapturous sound in *Meghe Dhaka Tara* (1960) also sparks strong emotional reactions. An obtrusive whiplash sound, used like a refrain in several passages, underscores the psychological state of the protagonist, Nita. The sound effect demarcates her inner universe from the rest of the film's narrative development, allowing it to resonate emotionally with the audience. In one sequence Nita, who will later die of tuberculosis after a long battle against her ill-fated social condition and sacrificing her life to keep her family together, is learning to sing a love song for a marriage ceremony within a melodramatic auditory setting. The ceremony refers to another tragic twist of Nita's fate, as her fiancé, Sanat, decides to marry her sister rather than her. The love song ends with the sound of repetitive whipping, complete with extended reverb. The unexpected, quasi-diegetic sound effect directs the audience's empathy towards the female protagonist, and the empathetic emotion takes the character as its 'intentional object' (Plantinga, 2009, p. 90). This approach to sound as a design element became majorly influential on the dominant melodramatic styles of Bombay-based Hindi cinema and on regional cinema grounded in situated realities such as local myths, customs, and rituals. This inclination towards mythic and ritual life makes Ghatak a more colloquial filmmaker than Ray, whose films sometimes appear to have 'export quality'.[4] Film scholar Erin O'Donnell elaborates on Ghatak's stylistic use of sound:

> The technical details of Ghatak's melodramatic style include the following stylistic traits: frequent use of a wide-angle lens, placement of the camera at very high, low and irregular angles, dramatic lighting composition, expressionistic acting style and experimentation with songs and sound effects. (O'Donnell, 2004)

Ghatak's experimentation with sound effects is widely considered as a merely melodramatic component of his work, but the modalities of this practice are not discussed enough. I find that sound operates as an essential element of narrative entanglements in his work. Actual and natural sounds take an autonomous position outside of their diegetic connection with the image. Such practice temporally expands the scope of the narrative to involve a collective consciousness of the communal, beyond the here and now. Ghatak's use of innovative techniques may be shocking, but it sensitizes audiences to the trauma of partition and migration. Film critic Shoma A. Chatterjee notes:

> His use of sound is not only aesthetic and imaginative but it is also startling, designed to shock, to reach beyond the cinematographic frame of the film. For him, sound could also be extremely effective even when it

4 Ray was criticized for his popularity in the West, and the explicit influences of European and American films on his work. Notable is popular actor Nargis Dutt's comment in the Indian parliament (1980), accusing Ray 'of distorting India's image abroad'. See: Andrew Robinson's biographical work on Ray, *The Inner Eye* (1989).

> is ornamental. *Meghe Dhaka Tara* is a classic example of the power of sound in Ghatak's cinema. ...soon after his marriage, Sanat tries to patch up with Nita. But Nita walks away in dignified silence. The sounds of the whiplash now torment Sanat for his betrayal of Nita. The use of Indian classical music (Shankar practices his *raga* out in the open), a Tagore song (sung by Shankar and Nita), and the *Baul* performance enhances the film's socio-historical and cultural context, making it specific to undivided Bengal, almost like a tribute to its cultural heritage. (2020)

The non-immersive/discursive aspect of Ghatak's work loosens the ties between the sited experience and subjective formation, sparking the spectator/listener's awareness of the protagonists' plight. Sound scholar Joanna Demers notes: 'Discursive accent, then, exists in a state of ambivalence, ... discursive accent resituates the phenomenal qualities of voice (or sound, taken broadly) into an artwork, and divests sound of its signifying properties so that it can conceal, rather than reveal meaning' (2013, p. 149). Drawing on her perspective, one can argue that a discursive situation in an artistic experience occurs when the spectator/listener is free to detach themselves from the experience. As a result, the use of sound opens the cinematic situation for multiple possible interpretations, rather than being fixed in an ontological relationship with the experience.

This disjuncture is crucial in my own sound art practice. It enables a self-aware, critical faculty to emerge through a personalized, post-immersive sonic experience. Inspired by Flusser's and Demers' ideas of a discursive accent, I have proposed three primary approaches (Chattopadhyay, 2020) to disrupt pleasant immersion and trigger this critical moment:

> 1) intrusion of noise, such as a sudden loud scream, as an alarm mechanism;
> 2) asynchronism – a divorce of sound and image in an audiovisual experience analogous to the asynchronous mode of cinema proposed by V. Pudovkin (1985, pp. 83–85);
> 3) disrupting the narrative flow with discursive elements such as critical commentaries, fragmented stream of thoughts, and poetic utterances.

In Ghatak's work all the above techniques are used in varying degrees to break the mode of immersion and sensitize audiences towards the plight of his protagonists, reflecting the injustices, crises, and complexities of the Indian condition of his time.

Such disjunctive moments in a narrative flow create a sense of what Gilles Deleuze and Félix Guattari (1972) term 'deterritorialization', a notion that resonates with Flusser's homelessness. Deleuze and Guattari's deterritorialization describes the disembedding and re-embedding of social relations from various objects and sites. The term describes the process that decontextualizes a set of relations, rendering them remote and virtual outside of the constraint of the enveloping here and now, and preparing them for further actualizations outside a fixed local territory. Many anthropologists use the term to refer to a 'weakening of ties' between culture and place, site and self, inferring the removal of cultural objects from a certain location in space and time for new subjectivities to form. In the context of a closed immersive medial experience, both the ideas of homelessness and deterritorialization help open up the experience to personalization and subjective intervention through this weakening of ties with the immersive space. Through this rupture, the individual spectator may regain their discursive faculty.

Ghatak's swansong, *Jukti Takko Aar Gappo* ('Reason, Debate, and a Story', 1974) uses the disjunctive moment to great effect. Although the film ends with a violent political encounter, the two surviving lovers, Bongobala and Nachiketa – the constant companions of nomadic protagonist Nilkantha (played by Ghatak himself) – communicate through the proximate sound of intense breathing. This quasi-diegetic underscoring of the sound, communicating love while on the battlefield, operates as a discursive sonic intervention and a symbolic antiwar message of peace and tenderness. The post-immersive/discursive approach in Ghatak's work is his political stance, a position that advocates for subjective awareness on contemporaneous socio-political issues in India at the expense of narrative pleasure. Ghatak's embodied practice of sound, alongside the use of intricate background music and songs, shows a new pathway of situated-ness in Indian culture, rituals, and myths, developing an indigenous syntax. In the hands of more recent filmmakers, these methodologies have expanded even further.

Mani Kaul and a Reflective Aurality

Mani Kaul was a student of Ghatak at FTII, Pune.[5] Like his tutor, Kaul didn't consider film as a purely storytelling medium. For him, cinema was a tool to explore the dynamics between sound and image, and cinematic

5 In later life Ghatak taught briefly at FTII, Pune, and had a strong impact on his students. See: www.cinemaazi.com/feature/-a-bottle-in-one-pocket-your-childhood-in-the-other-ritwik-ghatak-s-bombay-poona-years (accessed 1 March 2023).

time and space, drawing on Indian traditional arts and music, and working with the knowledge and craft embedded in indigenous ways of seeing and listening to throw light on the human condition. His primary fuel was Indian philosophy, ancient musical practices, and established schools of creative thought. In addition to following the Indian cinema canon during his formative years, which led him to study filmmaking, Kaul was trained in the oldest musical tradition in India, *Dhrupad*, by its foremost exponents, the senior Dagar Brothers.

Heavily inspired by the Dhrupad style of slow, contemplative musical improvizations to build fluid architectures of sound, Kaul deviated from the social realism of his time (e.g., the work of Ray and his followers) and focused his attention instead on the self-reflexive layering of a multitude of realities and the capacity for interior reflection. He complicated the then dominant sense of Indian cinematic time and space derived from western linear temporality and perspectival spatiality. Much like Dhrupad and other traditional Indian musical practices, his work constructs a reflective and indolent sense of space and recursive temporality. He used grainy vocal sounds and dubbing, at times out of sync with actors' lips, to realize clinically distanced dialogue. Sometimes muted, diegetic environmental sounds were also used. This experimentation with the scope of analogue sound aesthetics, technologies, and methodologies such as dubbing and post-synchronization worked in parallel with the conventions of Indian cinematic sound.

Kaul's subsequent interest in Robert Bresson's austere and minimalist style of narrative also infiltrated his use of sound. His films often deliberately broke cohesion between sound and image, altering the illusion of redundancy (Chion, 1994). This interest in minimalism is not exclusive to the European auteur; Kaul's work focused largely on abstraction, absence, and under-dramatization. In his first feature film *Uski Roti* ('Daily Bread', 1969), the director reflects on, rather than narrates, the predicament of a woman who walks every day to a bus stop to deliver bread to her husband. The film is an attempt to destroy any semblance of realism. Kaul constructed it like a piece of audio and visual architecture from units of movements and change, reflecting the methodology of a Dhrupad music performance in which a single unit of sound is treated like a building block. The resultant work reads as an artistic intervention improvized from many potential 'combinations of movements' (Kaul, 1977, p. 24). Kaul's next film, *Ashad Ka Ek Din* ('One Day in the Rainy Season', 1971), stripped its narration of all the drama. Throughout the film a static voice, almost unemotional, focuses on the literariness of the text.

This austere and abstract approach reached its zenith in his two films dedicated to Indian classical music, *Dhrupad* (1983) and *Siddheshwari* (1989). The former is a meditative and reflective commentary on the Dagar family of *Dhrupad* musicians and the latter a portrait of the legendary classical Indian music singer Siddheshwari Devi. Both films follow improvizational movements of classical music. Scholar Mantra Mukim notes: 'The analogy to music is crucial to Kaul because the process of listening comes very close to how he conceives of seeing within cinema – an act without definite and destined meaning, which is newly produced when an individual viewer encounters an individual moment in an individual scene' (2017, p. 90).

This generative and bottom-up approach to filmmaking involves the spectator/listener. As with India's traditional music and arts, Kaul wanted to engage his audience as a participant, and an empathic and responsive entity, rather than as a consumer of audiovisual media – an approach drawn from India's colonial roots and western influences. He wanted to challenge the idea of cinema as a pre-packaged entity delivered to its audience. The use of sound in his films, therefore, treats the audience as a co-listener, one who will lend their ears to multiple interpretations in an open-ended field of listening and sensing together.

This thrust on reciprocal communication in cinematic art was central to his sound practice. He wanted to convey that cinema should be free from its reliance on the visual image to embrace an exploration of temporality, whereby sound could act as a function of time. This obsession with a non-linear movement of time is reflected in all his works. In *Duvidha* ('Dilemma', 1973), past, present, and future times coalesce, and in *Nazar* ('The Gaze', 1991), memory plays a central role. *Smriti*, an Indian concept of memory, suggests the idea of the unknowable in human experience.[6] Kaul's aesthetic journey embraces this sense of abstraction, absence, and unknowing as vital forces in his work.

Kaul's films demand intelligence, requiring an audience to actively participate in the process of sensing and meaning-making. However, India's cinema audiences have been conditioned in consuming pre-packaged films since the colonial media of film arrived in South Asia. For them, Kaul was speaking in another language. They were not yet open enough to partake in a two-way conversation and so Kaul's work has been largely rejected. He died in 2011 without much recognition. However, Kaul's work requires

6 *Smrti* or *smriti* (Sanskrit for 'recollection') relates to human memory and knowledge transmission through remembrance and is defined as the capacity for retaining and sharing experiential knowledge.

book-length research given the enormity of his decolonial, aesthetic vision. Indeed, it is difficult to give full justice to his sound practice in a single subchapter. Kaul is one of India's most important, though highly underrated, filmmakers.

Shyam Benegal and Listening in Sync

Shyam Benegal is one of the most prominent next generation figures who partook in India's parallel cinema movement, drawing heavily on Ray-esque social realism.[7] Benegal's first film, *Ankur* ('The Seedling', 1974), is an early example of the Indian New Wave, much of which is marked by shooting on location with sync sound. In *Between the Headphones*, he discusses the film's use of sync sound:

> When I shot *Ankur* in 1973, I used a Seiki camera and it was entirely sync sound. I was among the first people to shoot a sync sound film in its entirety with a Seki camera. ...I took the cameraman who...had experience going back to the 1920's. I took a sound recordist who also had started his career in the 1930's. So I had two people who had immense experience with this kind of work. ...Then they used to dub their sounds in a studio. I refused to take a pilot track because at that time I was kind of fanatical about getting the right and exact sound. I was very much impressed with the idea, which was very current in Europe at that time – the use of direct sound. The whole idea was that you had a sense of reality that you got with it [direct sound] – impossible to get in the post-sync – the emotion, the performance. The performance was not broken into two parts – the visual performance and the sound performance. It was an integrated performance. I was absolutely the biggest advocate for that. When I made my film in 1973, I was among the very first people to do an entire sync sound film without taking recourse to any person dubbing. Not a single dialogue that you hear in *Ankur* was post recorded. Everything was done on the location where we shot the film, with natural sounds and all. Then we emphasized the

7 In an interview, Benegal describes how, as a young practitioner, he was mentored by Ray: the established filmmaker wrote him a letter of reference for a film fellowship; and, while developing his own cinematic language, Benegal was inspired by Ray's body of work and approach to cinema. See: https://thecolloquium2017.wordpress.com/2014/01/14/cinema-and-other-things-with-shyam-benegal/ (accessed 1 March 2023).

> soundtracks, had other tracks to add to the environmental sound and everything else. But there was nothing – no dialogue was ever post-sync recorded. It was all recorded at that time. Now, this is very important for me. It has always remained the most important thing when I make a film. (Chattopadhyay, 2021b, pp. 21–22)

Benegal describes the practice of using direct sound as a precursor to sync sound. In the fifth chapter, I discuss how Ray perfected direct sound methods to convey his cinematic vision of social realism and situatedness. In a typical Indian mainstream film of his time, sound was not meant to complement and equal the visuals in its narrative ability. In taking creative licence from Ray's work, Benegal challenges this redundant relationship by giving the soundtrack importance, endowing a situated listening in sync with the location and all its vibrant layers. This sonic situated-ness reveals India's socio-economic realities, from caste and gender inequalities to the psychological trauma of the Partition, feudalism, and post-colonial struggles. Listening in sync and the social situated-ness of the cinematic experience are central to his practice. Writer Gautam Pemmaraju recalls an anecdote revealing the extent to which this ethos of listening in sync unfolds in Benegal's work:

> Shyam Benegal's *Nishant* (1975) wherein Girish Karnad unsuccessfully resists the abduction of his wife, played by Shabana Azmi. As he walks around in agitation, pleading with the bystanders who watch on impassively, an eerie rhythmic chant occupies the soundtrack. The film was shot in Pochampally in Andhra Pradesh, and (Hitendra) Ghosh[8] recalls that as the crew were wrapping up the day's shoot in the 67s little town where darkness had already descended, they heard a strange, unsettling sound. Ghosh went off into the darkness to investigate, following and finally locating the sound: it was being chanted by a woman who had been visited by the devi, a 'possessed woman' who was in a trance. He unobtrusively made a recording. In the post-production stage, they found the background score falling somewhat short in expressing the helplessness and fear experienced by the emasculated husband. They decided to use the chant in lieu of music, as texture to convey the power of the scene. (2013, pp. 67–68)

8 Hitendra Ghosh worked as sound mixer for many of Benegal's films. I have also interviewed Ghosh in *Between the Headphones* (2021b), where he provided some other acecdotal insights on sync sound practice and the work of filmmakers, including Benegal.

This approach, engaging with the very site of cinematic observation and drawing narrative elements from the situation itself without extraneous means, is reflected in how music appears in Benegal's starkly presented diegetic worlds. Vanraj Bhatia composed music for most of Benegal's films, including *Ankur* (1974), *Nishant* ('Night's End', 1975), *Bhumika* ('The Role', 1977), *Kalyug* ('Age of Vice', 1981), and *Mandi* ('Market Place', 1983). Bhatia, along with other composers associated with the Parallel Cinema movement such as Vijay Raghav Rao and Rajat Dholakia, brought different kinds of sounds to their artistic contributions. Their film scores are restrained, thoughtful, and often atmospheric with a stronger and nuanced creative intervention when compared to the loud and frontal mounting of typically emotive and melodramatic background scores in mainstream Indian cinema. These parallel films also challenged the normative rules that song-and-dance routines were necessary for cinematic storytelling. Whereas popular films use songs as punctuational relief and overt entertainment for a mass audience, parallel films generally avoided them, because filmmakers like Benegal believed they would break the narrative flow; their cinematic intervention aimed at faithfully portraying India's post-independence socio-political reality. If songs were occasionally used, pieces were borrowed from Indian folk and vernacular musical forms. In Benegal's films alone, many regional musical forms influence the soundtracks such as local music from Gujarati in *Manthan* ('The Churning', 1976), the Marathi *Tamasha* performing arts tradition in *Bhumika* ('The Role', 1977), and North Indian folk music in *Junoon* ('Obsession', 1978). Film music often operates in collaboration with sound design in the work of Benegal and his contemporaries associated with the Parallel Cinema movement.

Other pioneers

Ray's legacy – exploring actual sounds to create realistic auditory settings, portraying the sites depicted in the story – has inspired independent social realist filmmakers like Adoor Gopalakrishnan in southern India, who also has a strong spiritual connection with Ritwik Ghatak's and Mrinal Sen's work. Gopalakrishnan is a towering figure in both Malayalam-language and Indian cinema. He has continued the tradition of direct or synchronized recording and monaural mixing, making subtle and naturalistic use of site-specific sounds. For Gopalakrishnan, sound is a vehicle for underlining certain situations and is used as a recurring theme. Each of his films is an exercise in the ornamental use of sounds that add psychological layers to the story being told. Gopalakrishnan deftly uses sounds such as incessant rain, a passing bird, crickets, forest murmurs, or busy traffic as part of the construction of film space and of the narration's mood. The director comments on his focus on preparing a 'sound script' prior to shooting:

> Sound is as important as the visual in films. I am very particular about the sound factor. In fact, I always have a separate script ready just for sound. ...I have uniquely used many sounds to convey a message or any idea. Only a close scrutiny will make you hear the sound. Like some scenes in *Mukhamukham*. (2009)

His use of natural and environmental sounds suggests parallels with character studies, location studies, or studies of certain situations. In the jail sequences of *Mathilukal* ('The Walls', 1990), a bird appears to suggest the desire for freedom, the sound of a cricket relates to a sense of loneliness, a running rabbit's call is understood as playful relief. Later in the film, a curious love relationship builds between the protagonist, Basheer, and Narayani, a woman who is imprisoned in the women's cell on the other side of the jail wall. She is never seen in the diegetic world, neither by the audience nor the protagonist, only an acousmatic voice marks her presence. The various emotive suggestions contained within the thrown, grainy voice make the characters' separated yet intimate conversation endearing and engaging.

Michel Chion introduced the idea of 'Acousmêtre' in *Audio-Vision* (1994), drawing on the work of composer Pierre Schaeffer. With this term, Chion studies the effect of sounds whose source or object remain unseen in cinema. For Chion, acousmatization endows a voice with an 'aura' of innocence. In its reversal, de-acousmatization, when the unseen speaker is revealed, 'the voice loses its virginal acousmatic powers, and re-enters the realm of human beings' (1994, p. 23). If we illuminate the voice of Narayani, Basheer's unseen lover in *Mathilukal*, with Chion's fertile terminology, her unvisualized sonic presence alleviates the protagonist/author's prison life and the cell's claustrophobia. Keeping her unseen, Gopalakrishnan grants this love affair an ethereal quality, a childlike innocence to transcend the trauma of incarceration.

G. Aravindan, another Malayalam filmmaker, also uses sound minimally, adding lyrical and poetical connotations. *Oridathu* (1987) examines the arrival of electricity in a rural area of Kerala. As discussed in *The Auditory Setting* (Chattopadhyay, 2021a), this sensitive film focuses on the cultural shock, resistance, and general reactions in the village to the slow and devastating processes of modernization and urbanization. This development slowly wipes out traditional and ritual sounds such as conch shells played in the evenings from the typical auditory setting of village households. They are replaced with the sounds of electrical lamp switches and lampposts. As crows are electrocuted, the village's ambience is absorbed by their fearful cries. The traditional acoustic ecology, therefore, begins to fall apart and is replaced with new sounds. Aravindan listened carefully to the transformation and his film has become a critique of technological modernity's effects that disturb intricate systems of tradition.

A dedicated student of Ghatak, Kumar Shahani made luminous and thoughtful films grounded in the Indian classical music and aesthesis; but he was also drawing from European films and their austere narration styles and reified (sonic) environments. After graduating from the Film and Television Institiute of India under the tuteldge of Ghatak, like Mani Kaul, later Shahani assisted Robert Bresson in two films traveling to France with a scholarship from the French Government. Nurtured both by Ghatak's ethos of Indian ritual and traditional consciousness and Bressonian minimalism, in his first film *Maya Darpan* ('The Illusory Mirror', 1972) East meets West in the formalist exploration of the cinematic medium and in the use of sound. Likewise, Shahani shows a remarkable narrative restrain, being interested rather in the symbolic uses of sound and the associative worlds they suggest. The recurrent sound of the free flocks of birds appears now and then in *Maya Darpan* and quite symbolically suggest the protagonist's rebellion against her father as a natural resistance of artists and freethinkers against traditional feudal systems. In the film, the discreet environmental and industrial sounds develop a contrapuntal relationship often symbolizing the contraditctions and tensions in post-indepence, modernizing Indian societies that the protagonists in the film are trying to negotiate. In *Tarang* ('Wages and Profits', Shahani, 1984) multiple voices are juxtaposed, some onscreen, some off-screen, all are equally striving for listening attention, symbolizing a future classless society given the many layers of class divisions existing in India. Shahani's use of sound is aesthetically grounded in Indian music, such as khayal, being careful toward its improvisational build up and aural ornamentations.

Saeed Akhtar Mirza turned to feature filmmaking after his work in advertisement films, and later studying at the Film and Television Instiiute of India and establishing himself as a documentary film practitioner. In his feature films, often with idiosyncratic but provocative titles and intriguing premises to begin with, urban middle class and their angst burst into the screen in a language of satire. Most of his characters struggle and search for identity in a rapidly changing Indian society and dwindling economic conditions of the

70s and 80s. The angst-ridden protagonists in *Albert Pinto Ko Gussa Kyon Aata Hai* ('Why Does Albert Pinto Get Angry?', 1980) and *Arvind Desai Ki Ajeeb Dastaan* ('The Strange Fate of Arvind Desai', 1978) speak in intensity, often in harsh words. Their voices are grainy, rising above the loud ambiences of the city. Announcements from radio and television inform the indoor settings of the home about the outside worlds marred with strikes, and unemployment. The typical medium-specific sounds of the magnetic recording from the 70s and 80s with monaural mixing render the voices and background musical scores high-pitched. Certain higher frequencies stand out in their shrillness due to the re-recording on optical 'married print' with their set compression and equalization. Amid such industry norms, the many layers of urban sounds are included and presented to accentuate the narratives of everyday struggles.

Girish Kasaravalli's films delve deeper into ritual practices and ceremonial religiosity ingrained in the Indian societies, particularly in southern India, in the state of Karnataka. He made his films mostly in Kannada, his native language. From his first film *Ghatashraddha* ('The Ritual', 1979), there manifests an interest in religiosity, dissecting the caste systems and social fragmentations via cinematic lense. As part of his method, he chose naturalistic sound design with a careful focus on locative ambient sounds to accompany or complement the re-recorded voices in the studio to ground his storytelling in the real sites of observation and commentary. Even in an outdoor setting of *Dweepa* ('Island', Kasaravalli, 2002), voices are re-recorded in the studio via dubbing. This combination of synchronous and asynchronous method adds to his storytelling devices.

Other filmmakers whose work enriches the Parallel Cinema movement include Buddhadeb Dasgupta, Goutam Ghose, and Rituparno Ghosh – all from Bengal. Dasgupta uses voice as a central sound component to take his narratives forward. However, his films are not vococentric. In many sequences of his first film, *Dooratwa* ('Distance', 1978), for example, we hear a contemplative sonic engagement with pro-filmic spaces and sites. Urban sounds of Kolkata have a crucial role in providing the setting for his characters. In his later films, *Bagh Bahadur* ('The Tiger Dancer', 1989), *Charachar* ('Shelter of the Wings', 1993a), *Tahader Katha* ('Their Story', 1993b), *Lal Darja* ('The Red Door', 1997), *Uttara* ('The Wrestlers', 2000), *Mondo Meyer Upakhyan* ('A Tale of a Naughty Girl', 2002), *Swapner Din* ('Chased by Dreams', 2005), and *Kaalpurush* ('Memories in the Mist', 2008), Dasgupta finds a fine balance between voice and other sound components, often site-specific sonic details illuminating his cinematic situations and their poetic suggestions. Anup Mukherjee, who recorded and designed the sound in *Mondo Meyer Upakhyan* (2002), describes Dasgupta's sensitivity and the spirit of experiment with sound:

> We didn't have the budget to do it in sync. It's not that it can't be done but who will do that? Most of the people fail to understand the basic requirements. Communication never happens properly. Anyway, what happened for that film was the director said that most of the situations in the film were outdoors, so if we would dub all that then it will look artificial. I asked him if he was sure and I would go ahead only if he was sure. So, we went ahead. He had a house in Mallickpur near Sonarpur. It is in a village. There was a huge lawn surrounding that house in the front and back as well. We packed the whole workstation along with the monitors and we dubbed most of it according to time slots (i.e., the day portions in the day and night ones in the night). We put up a tent and kept the front open and did dubbing. The dialogues inside the car were recorded by keeping the recorder in the bonnet of the car. When you see the film, you will realize that the audio is different, it is not regular audio. Not the usual crispness that we get. But still there is a field, there's open air – all that you will get along with the dialogues. When the characters speak, you will understand that it is under the open sky or it is inside a room. We arranged for another room inside the tent, which was indoor location, and the outside was the outdoor. This is how it was done. The footsteps were also recorded there. There was also a portion where the pushing of a car had to be recorded. It was more or less recorded that way. You can call this an experiment. (Chattopadhyay, 2021b, pp. 127–128)

Goutam Ghose's *Antarjali Jatra* ('The Voyage Beyond', 1987), *Padma Nadir Majhi* ('Boatman of the River Padma', 1992), and *Dekha* ('The Look', 2001) are examples in the restrained use of sound through many phases of technological transitions, since his career spans decades. In an interview Ghose elucidates how technological developments in Indian cinema have been negotiated by filmmakers:

> I really consider myself lucky to be a part of a lot of transition in the film industry. I started out with black and white 35mm film, worked on optical sound, used Michele and Arriflex cameras. Then magnetic sound came, then colour came and we then moved to the digital format. I have worked on every transition and transformation that hit the filmmaking workflow. If one thinks that the technology changed and they can't accept the new wave they can't compete in the game anymore. The market throws him off. So as filmmakers we have to constantly learn and adapt ourselves with every technological shift. (Nafiu, 2018)

Ghose used the conventional method of studio Foley in *Padma Nadir Majhi* (1992) as revealed by the film's sound mixer, the late Jyoti Chatterjee:

> When I joined NFDC, I worked for a film by Goutam Ghose. It's a film on Bangladesh called *Padma Nadir Majhi.* There the boat sounds, *boithar awaaj* (the sound of the water), *kaada* (footsteps on muck) – we created all these sounds by ourselves at the studio floor. (Chattopadhyay, 2021b, p. 37)

The director's experiments with Dolby SR in his later work *Dekha* (2001) heralded the digital era and the concept of 'sound design' in Bengali cinema:

> See, the word design in our country practically is happening post *Dekha.* Till this film there wasn't anything called a sound designer for a film. Whatever I am doing as a sound engineer, and Goutam Ghose as a director, we are sitting and deciding which sound will stay in which track, this is going to be used this way according to the script. That's how we will place these accordingly. Slowly this turned into design. (Chattopadhyay, 2021b, p. 121)

Despite the limited space of a book chapter, it would be amiss not to mention Rituparno Ghosh's films, too: notably, *Unishe April* ('April 19', 1994), *Bariwali* ('The Landlady', 1999), *Utsab* ('Festival', 2000), *Shubho Mahurat,* (2003), *Chokher Bali* ('Sand in the Eye', 2003), and *The Last Lear* (2007). His sound work presents a curious thirst for clarity in voice and post-synchronized aesthetics. The cleanliness of his films' diegetic universes is combined with dubbed voices, which often suspend the unfolding of a cinematic language in favour of an obtuse literariness.

The works of each of the abovementioned filmmakers merit book-length attention.

References

Carrigy, M. (2009a). "Bhuvan Shome". In Gopalan, L. (ed.), *The Cinema of India* (pp. 138–147). 24 Frames Series. London: Wallflower Press.

Carrigy, M. (2009b). "Meghe Dhaka Tara". In Gopalan, L. (ed.), *The Cinema of India* (pp. 126–137). 24 frames series. London: Wallflower Press.

Chatterjee, S. A. (2020). "Meghe Dhaka Tara". *Upperstall Magazine.* https://upperstall.com/film/meghe-dhaka-tara/ (accessed 4 July 2022).

Chattopadhyay, B. (2020). "Post-Immersion: Towards a Discursive Situation in Sound Art", *RUUKKU Studies in Artistic Research* 13. https://www.researchcatalogue.net/view/555994/555995 (accessed 10 September 2023).

Chattopadhyay, B. (2021a). *The Auditory Setting: Environmental Sounds in Film and Media Arts*. Edinburgh: Edinburgh University Press.

Chattopadhyay, B. (2021b). *Between the Headphones: Listening to the Practitioner* (2021b). Newcastle upon Tyne: Cambridge Scholars Publishing.

Chion, M. (1994). *Audio-Vision: Sound on Screen*. Translated and edited by C. Gorbman. New York: Columbia University Press.

Deleuze, G. and Guattari, F. (1972/2004). *Anti-Œdipus*. Translated by Robert Hurley, Mark Seem and Helen R. Lane. London and New York: Continuum.

Demers, J. (2013). "Discursive Accents in Some Recent Digital Media Works". In Vernallis, C., Herzog, A. and Richardson, J. (eds), *The Oxford Handbook of Sound and Image in Digital Media*, (pp. 140–153). Oxford: Oxford University Press.

Gopalakrishnan, A. (2009). "Before Resul Pookutty, there was Adoor Gopalakrishnan Rajaneesh Vilakudy". Rediff.com. www.rediff.com/movies/2009/mar/02adoor-gopalakrishnan-on-resul-pookutty-sound.htm (accessed 4 July 2022).

Kaul, M. (1977). "Communication". Symposium on THE CINEMA SITUATION, with Kumar Shahani, Adoor Gopalakrishnan, Shyam Benegal and Dileep Padgaonkar. https://theseventhart.files.wordpress.com/2011/07/the-cinema-situation_74.pdf (accessed 4 July 2022).

O'Donnell, E. (2004). "'Woman' and 'Homeland' in Ritwik Ghatak's Films: Constructing Post-Independence Bengali Cultural Identity". *Jump Cut: A Review of Contemporary Media* 47. www.ejumpcut.org/archive/jc47.2005/ghatak/text.html (accessed 4 July 2022).

Mukim, M. (2017). "The Private Eye: Mani Kaul and the Cinema to Come". *Caravan Magazine*.

Nafiu, R. F. (2018). "Goutam Ghose on the Future of Cinema". *Dhaka Tribune*. https://archive.dhakatribune.com/showtime/2018/04/20/goutam-ghose-future-cinema (accessed 4 July 2022).

Pemmaraju, G. (2013). "Soundbaazi: The Sound of More than Music". *ArtConnect: The IFA Magazine* 7(1), 62–82.

Plantinga, C. (2009). "Emotion and Affect". In *The Routledge Companion to Philosophy and Film*, edited by Carl R. Plantinga and Paisley Livingston (pp. 356–365). London: Routledge.

Robinson, A. (1989). *Satyajit Ray: The Inner Eye*. California: University of California Press.

Stubblefield, T. (2006). "Ritwik Ghatak and the Role of Sound in Representing Post-Partition Bengal". *Post Script; Commerce* 25(3), 17-29.

8. *Sholay*, Stereo Sound and the Auditory Spectacle

Abstract: Stereophonic mixing was introduced much later to Indian cinema than it was in Hollywood and had a relatively shorter lifespan. Following information on historical developments, critical analysis of sound in film and finding corresponding evidence in conversations with Indian sound practitioners, in the eighth chapter I demonstrate that stereophonic mixing, as an extension of magnetic recording and dubbing practices, rendered the cinematic imagination of this period as something spectacular; extravagant song-and-dance sequences shot in exotic locations and action scenes packed with studio-manipulated and synthetic sound effects which appealed to a mass audience.

Keywords: Indian cinema, magnetic tape, analogue media, stereophonic mixing, sonic space

Stereophonic Mixing

Although stereophonic mixing was feasible as early as the 1940s in Hollywood, film scholars James Buhler et al. recognize how, 'it would not be systematically exploited by the industry until the introduction of Cinemascope in the early 1950s' (Buhler, Neumeyer, and Deemer, 2010, p. 336). However, in Indian cinema, stereophonic mixing was introduced much later than in America and had a shorter lifespan; Dolby digital and its multichannel formats replaced multiple transitory stereo formats in just a few years.

Cinemascope was first used in India for *Kaagaz Ke Phool* (Dutt, 1959), but the film's sound mix was still mono.[1] Monaural mixing in the final mixdown for the projection and exhibition of films had been standard

1 See the film's technical specifications on the IMDb page: www.imdb.com/title/tt0052954/technical?ref_=tt_dt_spec (accessed 4 July 2022).

Chattopadhyay, B., *Sound in Indian Film and Audiovisual Media: History, Practices, and Production*. Taylor & Francis Group, 2024
DOI 10.5117/9789463724739_CH08

practice in India since the arrival of the talkies. Indian filmmakers indeed took quite some time to expand their sound palette. Raj Kapoor's *Around the World* (1967) would be one of India's earliest films to explore stereophonic mixing. Although a few filmmakers used the technology in the early 1970s, it wasn't until later in the decade that a series of films popularized the stereophonic sound experience. *Sholay* ('Embers', Sippy, 1975), in particular, marked a historic moment in Indian film sound largely due to these technical innovations.[2]

By the late 1970s, stereophonic sound technology came into full force when mainstream filmmakers like Prakash Mehra, Manmohan Desai, and Feroz Khan began using it in their work. What was the ramification of stereophonic technology's advent in cinema? According to film sound scholar Gianluca Sergi, stereophonic mixing made it possible to create an extra off-screen space in which part of the focus is left for the viewer to imagine. Stereo sound also made it possible for the viewer to engage with the sound's directionality. As the sound could move around and beyond the screen, stereo enhanced the audience's spatial orientation toward a fictional space (Sergi, 2004). Thus, unlike monophonic sound mixing, it became possible to make use of the off-screen space without having to link it to some visual on-screen reference. Sergi and other film sound scholars have pointed out the specific contribution of early stereophonic sound production to the expansion of filmic space, giving sound in cinema agency, a sense of autonomy, which nevertheless often compromised on sound nuances: 'Over the course of the 1950s, however, more and more films emphasized spectacle and grandeur, and the introduction of the widescreen format and stereo sound furthered this goal' (Buhler, Neumeyer, and Deemer, 2010, p. 339). This tendency is even more applicable to Indian cinema than Hollywood.

Sound as Spectacle: *Sholay* – A Case Study

Sholay, one of India's earliest prominent stereo films, was produced with 4-track stereophonic sound with spectacular effects. The film's combination of technological achievement and its substantial emotional appeal make it a reference point for both Hindi-language cinema audiences and the Indian film industry as a whole, even today. This mass appeal is exemplified in lines being repeated, the tone and texture of characters' voices being copied, and records of dialogue from the film, especially that of the villain,

2 See: trailer: www.youtube.com/watch?v=hLhzpe3_V_g (accessed 4 July 2022).

Gabbar Singh, still being popular with Indian audiences (Shankar, 2009). I have mentioned the way the specific sonic representation of the villain's character was constructed using vocal manipulation alongside extended reverb of his footsteps and other bodily postures and gestures. This narrative strategy, aimed at extracting visceral responses from the audience (Plantinga, 2009) triggered its mass appeal. The technically synthesized cinematic hyper-reality produced through studio-processed sound and stereophonic mixing centred on manipulating the audience's emotional engagement via the spatially expanded cinematic spectacle (Sergi, 2004). This narrative strategy, used to construct a spectacularly colourful and eventful fictional pro-filmic space, was slowly adopted in Indian cinema during the 1980s. Dissanayake and Sahai observe that:

> *Sholay* clearly is not a realistic film, there is very little social specificity inscribed in the filmic text. The narrative codes employed in the film serve to construct a metaphoric view of Indian society and its manifold problems. A metaphoric representation displaces accuracy. (Shankar, 2009, p. 165)

The synthetically reconstructed fictitious location (the village of Ramgarh), highly processed sound effects (e.g., scenes of brutal violence and elaborate fight sequences), and deliberately arranged folk-rhythmic song sequences intricately contribute to the film's intensity and emotional appeal (Shankar, 2009). The filmic world offers audience members a spectacular fantasy. Hitendra Ghosh, who assisted *Sholay*'s audiographer the late Mangesh Desai, says:

> The first time I used Foley was when *Sholay* came in 1975. *Sholay* was done in magnetic recorders that had four tracks. So, I suggested doing footsteps and all in one track. That's why you'll find that the footsteps of *Gabbar* on the stones in *Sholay*. Then there was a big thing – the coin that was flipped. Ramesh Sippy said, 'I want the sound of that coin.' At that time we said, 'Yeah, we'll do it.' When we were trying to get that sound by various ways of flipping coins, we were not getting anything. So we told him that there's no sound coming. He said, 'Oh! Don't worry. We'll tell R.D.' Then he told R.D. Barman to give the coin's sound. At that time synthesizers were new to the scene. They had just come out. R.D. was told to somehow use this tool and make that sound. He gave a lot of sounds and Ramesh Sippy rejected all of them. We were dubbing at Rajkamal then. So I used to wear kurta back in those days. I was coming down

> from the projection room and one coin fell down from my pocket. They have mosaic flooring and stairs. So, the coin fell and it made a certain sound. ...I heard it and I realized that it is a good sound. Why don't I do that? Then I got a big Electro-Voice mic from the dubbing theatre, we got the cable and put it there and then I threw one coin on the floor and recorded that sound. After recording, I went to editing. At that time we had Steenbeck. I saw the shot on it. Now the shot was different – the coin rolled down quite a bit. Now this sound was very small. So, what I did was, I cut the sound of the coin dropping on the mosaic and joined it. Even in the visual it was like this, a little bit of, you know, jerks. So this jerk and that jerk matched. ...I showed it to Rameshji. He said, 'Yes this is very good. But I guess a little more volume is required because there might be music here as well.' I said, 'Don't worry about the volume. That I can give you.' This mixing was done in Pinewood, London. ...We were all there. The recordist there in the mix said, 'What is this sound? The coin doesn't make any sound.' We said, 'No, we made this sir.' He said, 'No, I'll not do this.' He was so adamant that he would not use that sound, but Rameshji liked it. In the break, while having lunch, he asked me how we could convince that fellow. We insisted him a lot during the lunch. He said, 'No, I cannot use this. I'll be asked why. The coin doesn't make any sound. How did you use it?' I said, 'Just use it. We'll see what happens.' The whole day went in convincing him. Later he did it. After the film was completed, we didn't at all like the mix, which he did. So, we got the dialogue premix, effect premix and music premix from there and then Mangeshji mixed it in India at Rajkamal. That is what you hear now. We called him for the premiere. It happened at Minerva. I was sitting right next to him. When that coin sound came, the audience was stunned. As it was that any premiere show you went to at that time, people used to whistle, clap and all that used to happen. These days there are no such premiere shows. Public used to really go mad at that time. So, when they heard that sound of the coin in this sequence, they started shouting and applauding loudly. Then the mixer said, 'You people were right! Your audience is very different from ours.' I said, 'I told you, people will love this sound.' Even today people talk about that sound. (Chattopadhyay, 2021b, p. 142)

This anecdotal entry on the making of *Sholay*'s sound provides an insight into the processual development of expanding sonic spaces in Indian cinema, primarily contributed by young sound technicians in collaboration with directors, who were open to experiments and exploring the potential of sound to layer narrative. Spectacular sonic worlds, appropriated from India's mundane everyday life, created fantasy worlds of escape. The technological mediation of daily sounds into dramatic overtones was a tendency in stereophonic mixing during the dubbing era.

Sholay's practitioners recognized the quality of its sound design as a benchmark for mainstream Indian cinema's sound workers. Today's well-known sound designer Bishwadeep Chatterjee states: 'In mainstream, I think, one of the best sound design I have seen is in the film called *Sholay*. Watch that film over and over again. There's a lot of intelligent sound design that has gone into that film' (Chattopadhyay, 2021b, p. 434).

But he also notes that such innovations were not immediately recognized. Audiences still looked predominantly for the stars on screen and adapted slowly to stereo and later multichannel sound:

> Everybody sat up and noticed *Sholay*. It was never that the ambience track was not there. The ambience track was there. There were ambience tracks, there was this night cricket, there were the birds and all those things were there. It's just that we didn't know they were there. When people watched the film, they only knew the actor and the actress. They didn't know that there's a whole industry behind that. They didn't know that there was a director to start with, let alone the technicians. They went and saw the stars. (Chattopadhyay, 2021b, p. 435)

Sholay was one of the earliest Indian films that opened audiences to the potential of an expanded sonic space. Stereo settings augmented filmic narrative through a better developed aural domain. This expansion of the diegetic space often worked beyond the frame or the screen, which was a new experience in Indian cinema.

Despite *Sholay* ushering in a new era of potential spatial sound experiences, Indian stereophonic mixing continued along the road towards grandeur and spectacle, including works from the Bombay film industry and regional cinema in Kolkata, Chennai, and Kerala. The relatively higher dynamic range available to magnetic sound recording and the wider headroom in stereophonic mixing expanded the possibilities of studio-centric sound production. Placing music and sound effects on separate channels became common practice to spread the sound across the screen. Spectacular effects created an expanded fantasy-like experience, augmented with lavish songs and dances in foreign locations. Action sequences were packed with sound effects like echo and reverb, dislocating the sounds further from their realistic sources.

Expanded Sonic Space

The technological rivalry between several companies that followed saw the popular imaginary surrounding stereophonic sound in cinema intensify further. According to Indian film music scholar Rajiv Vijayakar, 'Polydor... made a mark in 1970, began Stereophonic sound with *Sholay* – since the film was also in Stereo – with *Julie*, *Chalte Chalte* and *Shalimar* to follow' (2009, p. 53). In this technologically stimulated, expanded sonic environment, more films embraced stereophonic sound to put on a big show for a demanding populace. During the late 1970s and 1980s, *Disco Dancer* (Subhash, 1982), *Gandhi* (Attenborough, 1982), *Mard* ('Macho', Desai, 1985), *Saagar* ('The Sea', Sippy, 1985), *Mr India* (Kapur, 1987), *Waqt ki Awaz* ('The Sound of Time', Bapayya, 1988) and *Maine Pyar Kiya* (Barjatya, 1989) all used stereophonic sound. The wider popularity of these films was linked to the use of stereophonic sound that 'foregrounded the spectacular experience' (Kerin, 2011, p. 28) rather than expanding creative practices. This populist focus, using spectacle and grandeur to cater for mass audiences and profits, was apparent in the lavish reviews some of these films received in film magazines and other popular media (*Hindustan Times*, 2010).

Sound practitioners, who at the time acted as the primary protagonists and interlocutors behind the developing sonic horizons of Indian cinema, were all enthusiastic about the possibility of expanding their sound pallet through stereo sound, even though producers pressured them to work overtime with little pay and directors demanded films were finished quickly. Veteran sound recordist and mixing engineer Anup Mukherjee talks about the trials and tribulations of early works made with six-track magnetic stereophonic mixing and the instability of production chains due to stereophony's brief timespan in India:

> Like a child's brain is sharper than an older person, the magnetic tape sounds great when it is new, but more it is run, its quality deteriorates. And after a certain period of use, a total loss of high frequency occurs. It is not that magnetic is the most satisfying format. In the age of 70mm, which was a six-track magnetic stereophonic production with 6 tracks on married print, films like *Lawrence of Arabia*, *Star Wars*, later *Sholay* in India, were made with six-track magnetic stereophonic mixing. The problem was that the print of these films never went bad, but audio did. Because there were different kinds of head used, let's say in India and in the US. So, the Ferrous Oxide coating would be removed slowly. It becomes drier. It catches oil that is applied to the projector, and it

> damages the tape. So, after each 30 runs you need to reprint the whole thing – only audio, not the visual. This is a problem. After final mixing, it's again back to square one, i.e. an optical track. As the final married print of the film is on an optical track, synchronized audio and visual. (Chattopadhyay, 2021b, p. 109)

Sound recordist Subash Sahoo, on the other hand, laments that *Sholay* was a lone phenomenon at the time of its making, and few films came close to its creative achievements. He goes on to state that the full potential of stereo sound wasn't as yet exploited:

> If you look at *Sholay*, it is one of the best sound design films of India, sound oriented film. But these kinds of films were less in number. Mostly they would accentuate music and dialogue. But as a sound person what I feel was happening in the optical medium was that our dynamic range was very less. It has the dynamic range of 78 KHz. It was theoretically 80-82, but practically 60 to something, whatever would go up to 6 or 7. Since the range was missing, the brightness was also not there. But it was correct for those times. But when the digital medium arrived, people used to record as per the frequency of the source. Then we replaced. Slowly, before Dolby, the Ultra Stereo came. But still *Sholay* was 70mm and 7.1. Magnetic medium was there during that time. Because of the magnetic and the optical medium, the dynamic range also was very less, though we are getting the perspective. I thought that particular format 7.1 wasn't used much because the 70mm print was rarely used. Of that, 7.1 wasn't actually successful at that time. (Chattopadhyay, 2021b, p. 311)

In *The Auditory Setting* (2021a), I discuss why monaural mixing was deemed inadequate for handling elaborate spatial information and sonic environments in film. Many filmmakers, especially those from American cinema, endeavoured to push technological limits and develop the craft of sound production to realize more ambitious film projects. Michel Chion locates a 'quiet revolution' (1991, pp. 69–80) in the intervention of Dolby, the company that made stereo mixes a regular possibility. India incorporated stereophonic mixing years later, largely due to a conservative take-up of new media technologies. The country's cinema and audiovisual media have long been known as prudent, taking fewer risks with frequent changes in technologies and aesthetics, whether introducing science fiction filmmaking or emerging technologies such as digital systems. The same happened with stereo sound.

Limited studio competency was met with issues regarding the changeover from monaural sound systems to stereo mixing and its reproduction in cinemas.

Disco Dancer (Subhash, 1982) and a handful of other films from the 1980s provide some of the earliest examples of stereophonic mixing in Indian cinema. A reviewer notes how 'a mixture of overextended musical sequences and overheated melodrama with action thriller overtones made it an over-the-top experience' (Clark, 1982). The expanded stereo sonic space, in this instance, created a spectacular musical exploration of the disco genre beyond the constraint of the screen. Encountering these kinds of populist experiences, Indian audiences began shifting their attitude towards film sound in stereo, from *listening to* it to *living in* it.

Even though stereo mixing didn't take off immediately, stereophonic sound gained some ground in Indian film song production and mixing in the 1970s. Innovative director V. Shantaram experimented with stereo mixing when recording the album for his film *Jal Bin Machhli Nritya Bin Bijli* ('Fish without Water, Lightning without Dance', 1971). However, Indian producers and film halls were slow to adapt to this new format, as they couldn't afford the new stereo mixing technologies that enabled sound to move from left to right with higher precision than monaural mixing. A cheaper alternative was found in 'echo' recording, or 'Electronically Reprocessed for Stereo', which was even explored within a few big budget films such as *Bobby* (Kapoor, 1973). Films whose songs were recorded in stereo include *Julie* (Sethumadhavan, 1975), scored by Rajesh Roshan and produced as a stereo LP. This trend of stereo LP record production continued for a brief period: a handful of such albums were released by HMV in the late 1970s, beginning with the vinyl release of the soundtrack from popular film *Amar Akbar Anthony* (Desai, 1977) and followed by *Karz* ('The Debt,' Ghai, 1980), produced in 4-track magnetic stereo sound. *Maine Pyar Kiya* (Barjatya, 1989) was a later film made in this advanced format. The method's economic unsustainability and technical limitations soon led to compact cassettes replacing records on the Indian soundtrack market.

In *Between the Headphones: Listening to the Practitioner* (Chattopadhyay, 2021b), veteran sound mixing engineer Anup Deb, who was one of the crew members of *Shahenshah* ('Emperor', Anand, 1988) and *Maine Pyar Kiya* (Barjatya, 1989), discusses how four-track stereo mixing and projection were developed. It was preceded by a system called two-track magnetic stereo that replaced the mono optical track, while the widely marketed Dolby two-track stereo mix was yet to come:

A: I was one of the team members of *Shahenshah*. We used to do the magnetic stereo at that time.
Q: Magnetic stereo?
A: That has got again left, centre, right, surround, but on the magnetic stripe.
Q: Was it separate from the film sprocket itself?
A: Yeah. One is outside the sprocket, another is the inside the sprocket. So this one side is two-track, another side is two-track.
Q: Okay. Magnetic stereo?
A: Magnetic stereo.
Q: Did that continue until this four-track?
A: Until Dolby came up with the – I should not say Dolby – Dolby came up with this thing long back, when Bollywood first got the optical stereo, that four-track stereo on the print.
Q: Which film, can you remember?
A: I think this was *1942: A Love Story*.
Q: That was the first Dolby four-track stereo, right?
A: It came here. Of course that film was mixed in Shepperton Studios, U.K. I've met those guys. They mixed there and for the first time we saw four-track optical stereo. You can't transfer on the negative and make a print because it is optical. ...That is the first time India got the print. It was mixed in London. Then, I think, *Rangeela* was the first film, which was mixed in India. Again, optical stereo.
Q: Optical stereo. Before it was magnetic stereo, was it separate from the film?
A: Magnetic stereo, yeah. Like the film *Maine Pyar Kiya* and all those you know, had magnetic stereo. Each reel we had to transfer individually and running time after checking whether the tape is correct, whether it is sounding correct. If there were any dropouts in the tape, your audio would get lost there.
Q: Hm.
A: I am the person who first mixed the Dolby Digital sound in India.
Q: Which film was it?
A: *Daud*. Ram Gopal Verma's *Daud*.
Q: *Daud*, yes.
A: That was the first Dolby Digital mix.
Q: How was the experience working on it?
A: That was my first film. I had an idea since I had worked earlier for the magnetic stereo. I tried to do my best, but of course things improve with the experience, you know.

> Q: Yeah. Could you put a little bit of dynamics right in sound?
> A: Yeah. Exactly. You now use the subwoofer, then use the surrounds, then keep the ambiences a little out – off-screen, you know, spread it and all those things. (Chattopadhyay, 2021b, p. 92–93)

The advent of stereophonic technology in cinema thus led to an emergent sense of spatiality in film as the sound palette extended, allowing actual spatial sounds to find importance in the scheme. Audiographer and mixing engineer Anup Mukherjee speaks about the hardships involved in innovatively expanding the sonic space in Indian cinema:

> See, during those times stereo mix on stereo opticals were done by a few guys. You need a stereo optical recorder to record in stereo and you need a stereo theatre as well. If I make it and no one else can hear – which is happening in a lot of cases – I'm giving highest quality output working on a Harrison mixer but the output is going through Ahuja amplifier or something like that, then what would you expect? Are we aware of what we should be doing? The chain of our whole production, from script to screen, do we know the required technicalities behind all this? And the kind of quality it should have? I think this is the main reason. With Tarun Majumdar's all in all encouragement I did a four-track stereo for his film *Path o Prasad*. That was magnetic track. How did I do it? I recorded one track on the outside of the perforation on a 35mm and another track inside in the place of the optical. Two such on either sides – one track between the sprocket and picture and one outside the sprocket and the same on either sides, which makes it four-tracks. They made this concept of four track stereo distribution. The playback was done on good projectors but the heads were four cassettes heads for each track. It was not their fault and the space was very less. After four rounds, the surface came off. That was what happened in reality. We never think about the end product. I've imagined and thought about a lot of times how to take any real picture of the Earth. I don't know how I'll do it. Maybe, on a rocket or a tabletop. Now there is DI and CG and everything seems to be possible with them. But imagine the hardship in those times, what Satyajit Ray imagined – imagine today's Goopy Gayen Bagha Bayen. How many kinds of magic did he create? What hardships did he go through to work on the optical? How many layers did we create? (Chattopadhyay, 2021b, p. 111)

In the dubbing era films, the relatively higher dynamic range available within magnetic sound recording and wider headroom in stereophonic mixing enlarged the possibilities of studio-centric sound production. Music and sound effects were placed on separate channels to spread the sound across the screen. This practice created spectacular effects for an expanded fantasy-like experience, augmented with lavish songs and dance routines in Indian films. Action sequences were packed with echo and reverb, modulating and engineering sounds even further from their realistic locative sources.

In the case of *Sholay* (Sippy, 1975), incorporating 4-track stereophonic sound mixing and re-production/projection created spectacular impact. The film achieved cult status both for its technological achievements and mass emotional appeal that this new spatial sonic experience had on the audience. The character's vocal manipulation, the extended reverb of his footsteps, other postures, and gestures, the synthetically reconstructed, fictitious location, the highly processed sound effects in brutally violent scenes and elaborate fight sequences, deliberately arranged song sequences – all intricately contribute to the film's technically synthesized hyper-reality, achieved through studio processing and stereophonic mixing, centred on manipulating the audience's emotional engagement with a spatially expanded cinematic sound experience. They engage audience members in the filmic world of a spectacular fantasy – a metaphoric and displaced representation of Indian society on celluloid.

References

Buhler, J., Neumeyer, D., and Deemer, R. (2010). *Hearing the Movies: Music and Sound in Film History*. New York: Oxford University Press.

Chattopadhyay, B. (2021a). *The Auditory Setting: Environmental Sounds in Film and Media Arts*. Edinburgh: Edinburgh University Press.

Chattopadhyay, B. (2021b). *Between the Headphones: Listening to the Practitioner* (2021b). Newcastle upon Tyne: Cambridge Scholars Publishing.

Chion, M. (1991). "Quiet Revolution...And Rigid Stagnation". Translated by Ben Brewster. *Rendering the Real* 58 (Autumn), 69-80.

Clark, G. (1982). "Review of 'Disco Dancer'". *The Spinning Image*. www.thespinningimage.co.uk/cultfilms/displaycultfilm.asp?reviewid=11568 (accessed 4 July 2022).

Kerins, M. (2011). *Beyond Dolby (Stereo): Cinema in the Digital Sound Age*. Bloomington: Indiana University Press.

Plantinga, C. (2009). "Emotion and Affect". In Plantinga, C. R. and Livingston, P. (eds) *The Routledge Companion to Philosophy and Film* (pp. 356–365). London: Routledge.

Sergi, G. (2004). *The Dolby Era: Film sound in Contemporary Hollywood*. Manchester: Manchester University Press.

Shankar, P. (2009). "Sholay". In Gopalan, L. (ed.), *The Cinema of India* (pp. 160–169). 24 Frames Series. London: Wallflower Press.

Times, H. (2010, July 24). "35 Years On, the Sholay Fire Still Burns". Hindustan Times. www.hindustantimes.com/bollywood/35-years-on-the-sholay-fire-still-burns/story-mcY7jLPKbKvYCwfrC9h4JP.html (accessed 4 July 2022).

Vijayakar, R. (2009). *The History of Indian Film Music: A Showcase of the Very Best in Hindi Cinema*. New Delhi: Times Group Books.

9. The Advent of Digital Sync Sound

Abstract: In the ninth chapter I study the advent of digital sync sound and the resultant (re-)emergence of spatial awareness in Indian cinema's 'digital era'. Since the late 1990s, a large-scale conversion from analogue recording and production practices to digital technologies has been taking place in Indian cinema. Digital technology has been integrated into the production and post-production stages of filmmaking, as well as to reproduction and projection formats. This chapter carefully studies how the processes of digitalization have made a substantial impact on the narrative strategies and aesthetic choices of extant cinematic sound production, informing the creation of spatial presence in film space via novel modes of interplay between quasi-diegesis and mimesis.

Keywords: Indian cinema, film history, digital technology, sync sound, spatial presence

The Auditory Setting in the Digital Era

It isn't difficult to observe that sound in Indian films has been inconsistently produced due to evolving phases of production practices, affected by technological innovations and shifts. There have been phases of sound practice (such as the entire 'dubbing era'), that cared less about quality, giving more importance to the typical narrative tropes such as song-and-dance sequences, dramatic use of sound effects, studio processing, sound editing, and dialogue replacement, as explained in chapter 6. However, there are also phases such as the 'digital era' (2000–) where some degree of careful and creative rendering of sound is observed. In this light, I consider the advent of digital technology in the form of 'sync sound'[1] and surround design in Indian cinema to be significant, given

1 I have previously unpacked this term: sync sound, which is a practical abbreviation of synchronized sound recording, meaning that the sound of voice and synchronized effects, as

Chattopadhyay, B., *Sound in Indian Film and Audiovisual Media: History, Practices, and Production.* Taylor & Francis Group, 2024
DOI 10.5117/9789463724739_CH09

the evolution of sound use through earlier phases of Indian cinema's production practices. The scholarly perspectives on sound production in the digital era in Hollywood (Kerins, 2011, 2006; Sergi, 2004; Sonnenschein, 2001) and the aesthetic impacts of digital production practices on the cinematic experience have often been described by film scholars as 'real' (Altman, 1992), 'sensorial' (Sobchack, 2005), or 'authentic' (Biancorosso, 2009). Although these expressions cannot be uniformly applied to Indian cinema in the context of its historical trajectory, much can be discussed along these lines. As I have mentioned in the Introduction and earlier chapters, mainstream Indian film's general tendency has been to ignore subtleties of sound while reconstructing the pro-filmic space into film space. These observations resonate with Indian film scholar Asish Rajadhyaksha when he points out the 'peculiar inability of Indian cinema to produce a persuasive relationship with live location sound' (2007, p. 1). He elaborates:

> To point to the inability of music to become sound, thus providing one context, and even a key explanation, for the peculiar inability of Indian cinema to produce a persuasive relationship with live location sound, the only proper sound resource actually available to the cinema...this in fact echoes the lament of all location recordists at the Indian cinema's curious resistance to live sound: both in the end questioning the dubious antecedents of the content of a film's soundtrack. (2007, p. 1)

The lament of the location sound recordist, however, tends to fade away with the emergence of contemporary digital tools and techniques that allow for the registering of multiple layers of sound, voice, effects, and ambient sounds directly from the location of shooting or from the film set. This includes the actor's performance and incorporation of these layers into the multi-channel environment of surround design. The result is a relatively believable presence of the voice of the characters as well as the pro-filmic space in the spatiotemporally constructed story-world that is achieved by using location sync sound as the primary element. In an interview, film producer Dipankar Chaki clarifies: 'I think the ambience is an extremely artistic aspect of the film sound, where it is probably one

well as a major portion of ambiences are recorded live on film set or on location during the actors performing their parts. This was a major development from the dubbing practices, in which all sounds were rerecorded in studio.

of the most behind-the-scenes things which is constantly colouring up[2]... the whole treatment of the film' (Chattopadhyay, 2021b, p. 524). How is the construction of the pro-filmic space in Indian cinema 'coloured' or affected by the use of sound in the narrative process of diegesis? Ashish Rajadhyaksha points out the overwhelming desire of Indian audiences to believe in the cinematic reality, which leads to a meaning-making in the mind of the audience with a delineated narrative based on conventional protocols of verisimilitude. He quotes Chion to support his point:

> The convergence – in which 'ambient sounds, which are often the product of multiple specific and local sources' do not recognize the hierarchy between a 'space inhabited by the sound' and its multisource origin – hinges on a confusion that is 'at the very heart of our experience itself, like an unsettled knot of problems'. This confusion has had significant technical consequences where the desire to read in the sound its origin has run counter to the conventional protocols of verisimilitude. (2009, p. 10)

The convergence thus expects the audience to delineate a story-world and make sense of a relatively more convincing presence of the characters and places narrated in the story using conventional protocols of verisimilitude. This sense of verisimilitude is provided by the qualitative attributes of sounds incorporated into digital film production: namely, textural richness, depth, perspective, volume, dynamic range, and the spatialization of on-location recordings. These qualities help the audience relate to the film space as a lived experience and as an extension of the phenomenal world (Bordwell, 2009).

The advent of digital technology in sound production (e.g., digital Hard Drive-based recording, live time-code-based synchronization with camera equipment, digital noise reduction tools, and platforms such as Dolby NR, multichannel surround sound design, and digital mixing) has overcome previous limitations, offering a wider and more flexible milieu of sound recording and design practices, and more freedom for the sound practitioner. Kerins writes of American cinema, which is also, to a certain degree, a valid comment in relation to Indian cinema:

> When 5.1-channel digital surround sound (DSS) first appeared in the early 1990s, it offered filmmakers better dynamic range, more channels,

2 This perspective on 'colouring up' with ambience resonates with Brian Eno's conceptualization of ambience as 'a tint'. See: 'Music for Airports Liner Notes' (Eno, 1978) http://music.hyperreal.org/artists/brian_eno/MFA-txt.html.

> and greater flexibility for placement of sounds within the multichannel environment. (2011, p. 53)

As previously discussed, the 35mm analogue optical filmstrip had a dynamic range of 78 dB (Kerins, 2011, p. 54) limiting the signal-to-noise ratio in optical sound recording. Within this narrower dynamic range, monaural synchronized recording was restricted in terms of the amount of sound content that would be optimal for a film soundtrack, thus placing emphasis on the voice. In the magnetic recording era, the dynamic range of sound recording was around 98 dB, depending on the quality of tape material. The digital format, on the other hand, offers more than 100 dB of dynamic range, which means that sounds can include more breadth and depth of recording (i.e., retaining high volume capabilities alongside the transmission of very soft and minute sounds). This wider digital headroom, with its inclusive capacity for recording, layering, designing, and mixing sounds, has allowed the gradual replacement of dubbing practices, Foley, and stock ambient sounds with more of the actor's live performance, sync sound effects, and location-based environmental sounds. When these 'actual' sounds are experienced by the audience through a spatially wider, digitally cleaned surround design by film-school-educated sound designers it can trigger a subjective sense of presence (of the pro-filmic space and the characters in it) in the auditory perception and cognition of the listener, leading to an embodied experience of sound. This mode of production inherently engages with sound's corporeally immersive potential. Mark Grimshaw argues that the use of digital synchronization of sounds in audiovisual media 'can create a sense of physical presence' (2011, p. 38). This immersive sense of presence in the cinematic universe is further consolidated with the surround spatialization of site-specific sound recorded on location, synchronized with the filmic action and characters' performance, and spatially reorganized on digital workstations. These digitally recorded and produced sounds have the capacity to envelop the audience often beyond the cinematic screen yet diegetically connected to the story-world, producing a perception of 'being there' (Chattopadhyay, 2016). This process of mimesis brings into play a 'coherent representation of the sound world' (McAdams and Bigand, 1993) in the spatially reorganized cinematic sound experience.

Digital Sync Sound: The Case of *Lagaan*

The first mainstream Indian film that was shot mainly with digital multi-track sync sound was *Lagaan* ('Taxation', Gowariker, 2001). As discussed in chapter 7, sync sound had already been explored by Parallel Cinema filmmakers such as Shyam Benegal. But *Lagaan* legitimized this practice and the profound ramifications of using situated sound in the Indian popular imagination for the first time. In this film, location sync recording and Dolby digital sound technology were implemented following a major debate in the Indian film industry.[3] And since then, most filmmakers have gradually embraced the so-called digital revolution. This is why *Lagaan* represents an important benchmark in the history of Indian cinema: shot entirely on a difficult yet quintessentially Indian location using sync sound recording, this film prompted recognition of the film location and its sonic environment as a significant actor in its story development. *Lagaan* was a trendsetter in its celebration of the sync sound technique, in large part due to sound technician Nakul Kamte's artistry and skill.[4] This qualitative shift has become evident in the way the film industry's crews, actors, and directors have adapted to the changing circumstances and how these changes have been reflected in their work, with consideration for commercial viability and functionality of sync sound.

Some relevant comments from prominent industry members, published in an issue of *Upperstall Magazine* (2002) focused on sync sound and Indian cinema, shed more light on this shift. Producer and filmstar Aamir Khan, who worked on *Lagaan*, discusses the advantages of sync sound over dubbed sound in transmitting a more natural actor performance that cannot be recreated inside the studio:

> I believe it is most favourable for an artist, as it enhances their performance and they can successfully record both emotions: sound and mime, and also avoid unattainable repeating of sentiment in a vacuumed dubbing studio. Certainly 'sync' sound will become a preferred way of working especially amongst artists, as it directly results in an enhanced performance. (2002)

Veteran actor Om Puri echoes Amir Khan's endorsements of sync sound recording experienced from in front of the camera:

> [F]rom an actor's perspective, dubbing is tedious and un-attached to their performance on the screen. Sync Sound helps sustain a coherent and natural presentation. (2002)

3 This debate inspired a number of public forums and publications, including an issue of *Upperstall Magazine* entitled 'Sync Sound and Indian Cinema' (2002).

4 For an elaborate description of the sync sound methods used in *Lagaan*, read the interview with Nakul Kamte in *Between the Headphones: Listening to the Practitioner* (Chattopadhyay, 2021b).

Noted filmmaker Nagesh Kukunoor, who utilized sync sound in his debut film *Hyderabad Blues* (1998), supports the use of sync sound in Indian cinema for aiding professional practice and film quality:

> Sync Sound is not at all limiting. In fact quite the contrary: it liberates the performance of the artist. Definitely, it helps promote increased professionalism and discipline on the sets, which is usually absent in our film industry. Had I my way, I would prefer using Sync Sound in all my films. This technique can certainly be promoted if the Director and Producer make a call to understand that Sync Sound is as good as, if not better than dubbed sound. (2002)

Independent filmmaker Dev Benegal comments on the 'lifelikeness' of sync sound and reiterates that the practice can facilitate an improved filmmaking process:

> Sync Sound breathes life into a film. ...Sync Sound unites the cast and crew and facilitates smooth functioning. I would urge all filmmakers to begin using this method, which is actually an untapped gold mine. (2002)

These sound technicians and practitioners, who have embraced the shift and appreciate the flexibility and creative freedom involved with sync sound more than other film crews, express their enthusiasm by fully implementing the technique within their practice. Renowned sound recording and mixing engineer Manas Choudhury, who I interviewed (Chattopadhyay, 2021b), and who also contributed to *Upperstall Magazine*'s issue on sync sound (2002), spoke in the magazine of the hierarchical relationship between sound and camera personnel in Indian cinema and sheds light on the larger debate concerning the domination of visual over sonic elements (2002). Further, in *Between the Headphones* (2021b), he stressed that the creativity involved in sound practice following the advent of digital sync recording and surround sound design upsets this hierarchy to situate sound in a more creative and innovative context.

Another well-known sound recordist Ashwin Balsavar shares a similar view on digital technology, considering that its higher quality recording and audio workstations provide more efficient techniques, facilitating honed skills and creative practices with sound (2002). Moreover, sound recordists within digital sync sound film production acquire equal importance as camerapersons and other technical stakeholders, leading to more appropriate cooperation and equality within the film crew.[5] Balsavar notes:

> Sync Sound texture is very realistic and cannot be reproduced in a Dubbing Studio and the actor's performances can never be repeated while dubbing either. Dubbing anyway never gives 100 percent Lip Sync. Dolby Digital, Non-linear editing, better microphones and Audio Workstations provide efficient technology and facilitate better postproduction. ...The final output in the theatre makes the film look and feel real. ...Dubbing is time consuming, patience testing and is an exercise in futility. Sync sound saves time and sounds better and gives due respect to the sound people. Big banners should adopt the use of 'sync' sound and make it a norm. (2002)

Balsavar's comments focus on the quality of digital multi-track sync sound. According to him, sync sound texture is 'realistic' compared with dubbing, which he considers never provided complete lip-sync – an opinion that I share.

Sounding the Sync

Given this overwhelming support for sync sound, *Lagaan*,[6] alongside other prominent mainstream Indian films made with sync sound since 2001, offers authentic sound layers and spatial textures that were previously unheard of in analogue mono or analogue stereophonic renderings of sound components in Indian cinema. The opening sequence draws the audience into the universe of the historic region of Champaner in 1890 via a 'real' or 'lifelike' sonic experience and perspective. Set in the ancient village of Kanuria, located a few miles away from Bhuj in Gujarat's Kutch district, the sound crew located the actors in the dry, empty, and hilly landscape of the village to record a major portion of the sound effects, voices, and ambiences in sync with the camera.[7] This wide and rustic Indian landscape comes alive with hitherto unheard clarity and details in the use of sync sound. When the

5 The emergence of the creative skillset includes the specialization of certain production processes, such as sound editing, production mixing, sound design, and location recording. These practices were previously taken up by one or two technicians.

6 Official trailer of the film: www.youtube.com/watch?v=oSIGQoYkFxs (accessed 4 July 2022).

7 See the IMDb entries on the film for further information on the film's site of recording and the pro-filmic spaces: www.imdb.com/title/tt0169102/trivia (accessed 4 July 2022).

characters speak, the voices carry a plausible level of aural reflections from the various architectural surfaces of the village – be it indoors or outdoors, in a field or inside a temple courtyard on the hills. These reflections were new additions to the sonic palette of Indian cinema, adding a realistic layer of spatiality and believability.

Post *Lagaan*, a number of films were released that primarily use sync sound such as *Dil Chahta Hai* ('The Heart Desires', Akhtar, 2001), *Swades* ('Homeland', Gowariker, 2004), and *Rock On!!* (Kapoor, 2008). Among these early, mainstream sync sound films, *Dil Chahta Hai* enjoys cult status thanks to its refreshing attitude, embracing the 'here and now' approach provided largely by the practice of shooting on location with sync recording, followed by surround sound design. Right from the beginning of the film, which is set in the contemporary city of Mumbai, the accurate sound portrayal of the traffic was in such contrast to audience expectations and experiences with studio-recorded sound that it placed them immediately 'on the streets'. A major portion of the sounds used in the indoor sequences consists of room tone, a noise-like 'hum' (Holman, 2002) coming from different electrical, electronic, and other indoor machines such as air conditioners.[8] The incorporation of such room tones in the ambience layer of film sound organization was a novel approach at the time. Each indoor sequence thus affords the right perspective and placement of the characters within the mise-en-sonore, which is heard not only through the reflection of voices on walls but also through the room tone surrounding the audience. In a major climactic sequence of the film,[9] the synchronized sound perspective manifests in the diegetic use of wedding songs sung outdoors at the ritual of a forthcoming marriage ceremony. This sequence was noteworthy for its novel use of diegetic music without any studio processing; the overly dramatic overtones of earlier Indian films, typically using delay and reverb, had been replaced with believable sound. By using sync sound at the location of the story-world, the sequence offers ample cues to the audiences about the pro-filmic space, which is sonically reconstructed to the point that the audience feels immersed in the embodied experience of its sonic presence.[10]

Swades (2004) offers a departure from the many post-millennial Indian films that follow a conventional formula narrating the global, diasporic

8 In a recent article 'Room Ambience: Home as Heard in Film and Media Arts' (2022b), I discuss the use and role of room tone in a number of films, including *Dil Chahta Hai*, that provides a sense of indoor presence.

9 Link to the film: www.dailymotion.com/video/x2j637i (accessed 4 July 2022).

10 See: interview with Nakul Kamte, Sound Designer of *Dil Chahta Hai*, shedding light on sync sound methods in *Between the Headphones: Listening to the Practitioner* (Chattopadhyay, 2021b).

experience of moving abroad by pursuing a cinematic return: a restored gaze towards the native land. It is a delicate and sensitive film about the nomad's homecoming. Describing the conditions of an expatriate Indian scientist in the US, the film traces the journey of the protagonist to his homeland through a process of self-discovery. In one decisive sequence, the protagonist sets out to visit a distant village to collect tax for a lease. The journey proves to be an 'ear-opening' experience for him. He meets the poverty-stricken villagers in this remote and rustic land while he accepts the kind hospitality of local hosts. Set inside his host's dilapidated hut, the protagonist's encounters with the embodied realities of Indian village life unfold through a sensitive and attentive listening to the voices and acoustic spaces that remain in the marginal territories of Indian society. The family representative narrates the condition of their predicaments: agricultural crops were all destroyed that year due to lack of water and, being of a lower caste, their family had faced the profoundly unjust distribution of water to their land. He breaks down while narrating this plight, his voice plausibly reflecting from the inner walls of his mud-built hut, carrying the sonic environment of this austere and subliminal auditory setting. As the audience, we hear the hesitant pauses in his speech as it should sound in a sync sound recording, situated in a rural heartland of India replete with adverse conditions, without any hint of studio processing or musical support. The use of sync sound thus transcends the synthetic making of sound, ushering in a new era of sonic actuality, turning the ear to the very site of locative existence – the earth and the everyday struggle of people – rather than a sonic escape from the 'here and now'.

The use of sync sound and surround design reaches a technical high point in the acclaimed, award-winning film *Slumdog Millionaire* (Boyle, 2008),[11] an essentially Indian production due to the number of actors, writers, locations in India, and technicians from the Indian film industry used. The film's production sound mixer and location sound recordist, Resul Pookutty, won an academy award for his sound work (Reuters, 2009). He later became one of the flag-bearers for sync sound in Indian cinema. In this film, several sequences are shot on location in slum areas of Mumbai using multi-track digital sync recording.[12] These sequences portray the complex depth of acoustic environments that Indian urban sites offer. Right from the

11 See: excerpts from *Slumdog Millionaire:* www.youtube.com/watch?v=QII_TAJ2BBY (accessed 4 July 2022).

12 See: interview with Resul Pookutty in *Between the Headphones: Listening to the Practitioner* (Chattopadhyay, 2021b).

opening sequence in the police station with its rich, multi-layered rendering of environmental sounds, found and recorded in Mumbai city, the audience is provided with bodily sensations of the site's spatial presence (Skalski and Whitbred, 2010). For example, the sync recordings of the actors' voices carry the dense ambience and detailed acoustic reflections of the room and the site of actuality that they occupy. Added to this are the respective room tones and noises in the spatially elaborate auditory settings. Outdoor sequences in Mumbai slum areas come to life with the immersive quality of ambient sounds recorded in digital multi-track sync technique at the locations where the story unfolds. In one sequence, two estranged brothers meet in a housing estate under construction. The background clamor of the construction site, including the drilling machine (obviously recorded in sync sound), provides the perfect setting for their meeting; the dynamics of their relationship is reconstructed as an intense sonic battle ground in readiness for the upcoming narrative threads: abduction, incarceration, and the struggle for emancipation.

In the subsequent phase of Indian cinema, the practice and use of sync sound gained momentum as more films employed this production practice. 'Independent' filmmakers, who prefer to stand apart from mainstream Indian film production to establish auteurist signatures of their own, picked up on sync sound as a stylistic feature of their work. Dibakar Banerjee, among others from this new breed of Indian filmmaker, used location sync sound to its fullest potential. In *Shanghai* (Banerjee, 2012),[13] the raw, rustic 'soundscape'[14] of an Indian city and its familiar phenomenal world is represented 'true to life' through sync sound and surround design.[15] Let us go through some of the substantial reviews of the film to illustrate the point I am making. Raja Sen of *Rediff.com* (2012) points out the sense of familiarity evoked by the sound strategy: 'The time is now, the location pointedly fictional and decidedly familiar.' Saibal Chatterjee of *NDTV Movies.com* (2012a) stresses the sense of place provided by sound: 'Lensed with great sense of place and occasion…*Shanghai* projects the dark, dank, redolent-with-danger innards of small-town India to absolute perfection…The most

13 See theatrical trailer: www.youtube.com/watch?v=ozitAKOrJVU (accessed 4 July 2022).

14 Here the notion of 'soundscape' (Schafer, 1994; Drever, 2002) is considered a point of departure for studying the evocation of the site in the sound of Indian cinema as a shift away from the linear film 'soundtrack' despite the term soundscape having limitations in understanding the complex relationship between sound, site, formation of space, and spatiality. I have critically reassessed the notion of soundscape in *The Auditory Setting* (2021a).

15 See: interview with sound designer Pritam Das in *Between the Headphones: Listening to the Practitioner* Chattopadhyay, 2021b).

striking aspect of Shanghai is its marvelous use of sound, both ambient and otherwise, to build up dramatic tension.' Chatterjee in *The Sunday Indian* (2012b) emphasizes the atmospheric and convincing use of ambience: 'They embrace the ambience of Shanghai with complete conviction, aiding and abetting the build-up of tension and atmosphere.' Hitendra Ghosh (2012), *Shanghai*'s re-recordist, confirms these claims: he explains how surround sound was used to create 'a very real experience…immersing the audience into the director's narrative'. He points to the novelty of the approach and the potential shift in contemporary Indian cinema: '[F]or the first time you will notice throughout the film that we have not used much of the foley [sic] sounds recorded in the studio. We have tried to use sound from the location. That's why the feature sounds very real and authentic' (2012).

The mainstream Hindi-language road-movie *Highway* (Ali, 2014) is at heart a metaphysical journey of self-discovery, a coming-of-age story set against the backdrop of evocative yet passing locations of northern India. Despite its initial premise to remain a typical road movie, the film manages to make several social commentaries about the sexual abuse of children in family settings, the profound social and economic inequality in Indian society, and the false hollowness of the Indian bourgeoisie. The running road is in the foreground, setting the narrative development. The characters, often displaying idiosyncratic movements, behave unpredictably given the outdoor freedom they start to enjoy. The film comes from an embodied understanding of India's heartland and the scenic beauty it possesses. The two protagonists – an abducted girl and one of her fugitive captors – travel through northern India in a truck, hiding in several places for a few days before running away. Sync sound methods allowed the sound recordist and designer Resul Pookutty to trace these passing sites and draw from their soundmarks. One reviewer on *Rediff* (Chhabra, 2014) notes the film's authentic portrayal of locations: 'The film seems to have so much energy, heart and authenticity – from the truck out on the highways, roadside dhabas and shops.'

It is only sync sound that could have achieved this sense of 'authenticity'. However, the compulsion to achieve clarity in film sound often leads the practitioners to use selected 'soundmarks' instead of faithfully capturing the thorough ambience of these sites. Each place is established with a certain 'soundmark' specific to the site. A location in Rajasthan, for example, is narrated through the distant and proximate calls of peacocks, since the state is well known for its variety of peafowl. This tendency to underline a particular sound, often at the expense of many other ambient sounds emanating from specific sites, serves as a kind of sonic 'compensation' for the noise reduction and editing of digital multi-track sync sounds in post-production. These 'industrial' norms, rules and regulations embedded in the film industry's sound production hinders the sound practitioner from using a more artistic approach of sited and situated listening that might further enrich the spatial experience offered by the film.

Despite these limitations, *Highway* stands out for its creative use of sync sound techniques, underscoring certain messages by stripping some pivotal sequences of any distractions to focus squarely on the fundamental issue in hand, like a keynote. One such sequence comes towards the end of the film which triggers a visceral reaction from the audience that has been compassionately sensitized to the issue of child abuse, especially of girls in familial settings. Veera, the 'abducted' girl, finally confronts her tormentor at home and allows herself to feel the pain caused by the loss of love she has suffered due to social inequalities. The primary buildup of this sequence happens by focussing on the grain of the voice, recorded in sync sound indoors. The heavy room tone, consisting of consumer electronics like air conditioning, adds to the tension. The banal voices of the familial gathering fall helplessly on the interior walls insulated from the outside world, reflecting back in hollow overtones, creating a sense of claustrophobia. Veera's voice, meanwhile, takes centre-stage, uttering unspoken truths for the first time. We hear every breath and deliberate pause in her speech, and the grunts and screams of angst and internal resistance from the girl who is now empowered to speak after experiencing love and the freedom of the outdoors. Veera's freedom and empowerment, enacted through her reclaimed voice, allows the audiences to empathize with her and with themselves in this cathartic scene. The provocative auditory setting pushes the audience to confront their own struggles with abuse, trauma, and the shame of not speaking out in time. It is a scene helped by sync sound technique, creating locative listening; the sonic spatiality and auditory situatedness accentuate the restrained drama of the film's closing moments.

References

Altman, R. (ed.) (1992). *Sound Theory/Sound Practice*. New York: Routledge.

Biancorosso, G. (2009). "Sound". In Plantinga, C. R. and Livingston, P. (eds) *The Routledge Companion to Philosophy and Film* (pp. 260–267). London: Routledge.

Bordwell, D. (2009). "Cognitive Theory". In Livingston, P. & Plantinga, C. (eds), *The Routledge Companion to Philosophy and Film* (pp. 356–365). London: Routledge.

Chatterjee, S. (2012a, June 7). "Movie Review Shanghai". *NDTV Movies.com.* http://movies.ndtv.com/movie-reviews/movie-review-shanghai-715 (accessed 4 July 2022).

Chatterjee, S. (2012b, June 8). "Movie Review Shanghai". *The Sunday Indian.* www.thesundayindian.com/en/story/movie-review-shanghai/112/36105/ (accessed 4 July 2022).

Chattopadhyay, B. (2016). "Being There: Evocation of the Site in Contemporary Indian Cinema". *Journal of Sonic Studies* 12. https://www.researchcatalogue.net/view/286592/286593 (accessed 10 September 2023).

Chattopadhyay, B. (2021b). *Between the Headphones: Listening to the Practitioner* (2021b). Newcastle upon Tyne: Cambridge Scholars Publishing.

Chhabra, A. (2014, February 14). "Review: Highway is Extremely Disappointing". *Rediff.com*,. www.rediff.com/movies/report/review-highway-is-extremely-disappointing/20140214.htm (accessed 4 July 2022).

Ghosh, H. (2012). "Notes on the Use of Sound in *Shanghai*". FutureWork. https://m.facebook.com/note.php?note_id=382123985181336&_ft_ (accessed 4 July 2022).

Holman, T. (2002). *Sound for Film and Television.* Boston: Focal Press.

Kerins, M. (2011). *Beyond Dolby (Stereo): Cinema in the Digital Sound Age.* Bloomington: Indiana University Press.

Kerins, M. (2006). "Narration in the Cinema of Digital Sound". *The Velvet Light Trap* 58 (Fall), 41–54.

Khan, A., Puri, O., Benegal, D. Choudhury, M., Balsavar, A. et al (2002). "Sync Sound and Indian Cinema". *Upperstall Magazine.* https://budhaditya.org/wp-content/uploads/2016/07/Sync-Sound-and-Indian-Cinema-Upperstall.Com_1.pdf (accessed 5 July 2023).

McAdams, S. & Bigand, E. (1993). *Thinking in Sound: The Cognitive Psychology of Human Audition.* New York: Oxford University Press.

Rajadhyaksha, A. (2009). *Indian Cinema in the Time of Celluloid: From Bollywood to the Emergency.* New Delhi: Indiana University Press.

Rajadhyaksha, A. (2007). "An Aesthetic for Film Sound in India?" *Journal of the Moving Image* 6. https://www.jmionline.org/article/an_aesthetic_for_film_sound_in_india (accessed 7 September 2023).

Reuters (2009, February 23). "Resul Pookutty wins Sound Mixing Oscar for 'Slumdog'". www.reuters.com/article/idINIndia-38161420090223 (accessed 4 July 2022).

Sen, R. (2012, June 8). "Review: Shanghai, a Frighteningly Fine Film." *Rediff.com.* www.rediff.com/movies/review/review-shanghai-a-frighteningly-finefilm/20120608.htm (accessed 4 July 2022).

Sergi, G. (2004). *The Dolby Era: Film Sound in Contemporary Hollywood.* Manchester: Manchester University Press.

Skalski, P. & Whitbred, R. (2010). "Image versus Sound: A Comparison of Formal Feature Effects on Presence and Video Game Enjoyment". *PsychNology Journal* 8(1), 67–84.

Sobchack, V. (2005). "When the Ear Dreams: Dolby Digital and the Imagination of Sound". *Film Quarterly* 58(4), 2–15.

Sonnenschein, D. (2001). *Sound Design: The Expressive Power of Music, Voice, and Sound Effects in Cinema*. Studio City, CA: Michael Wiese Productions.

10. The Surround Revolution

Abstract: The tenth chapter expands the discussion of the digital realm of sound production in Indian cinema towards the ensuing surround sound revolution. The spatial aspect of surround sound mixing and reproduction is becoming a key element in contemporary film production. As interest in Indian cinema flourishes via a healthy distribution market, both locally and globally, fostered by OTT platforms, the implementation of spatial audio innovations such as Dolby Atmos needs to be upheld to maintain this resurgent audience. In this chapter, representative Indian films that have been using spatial audio from 5.1 to the present era of Dolby Atmos are discussed to study their mimetic approach to site-specificity and the spatial awareness such films encourage.

Keywords: surround sound, Indian cinema, digital technology, Dolby Atmos, OTT platforms

The Advent of Surround Sound

In the 1970s and 1980s, Dolby Labs in the USA, run by Ray Dolby, improved the cinematic sound experience. Noise reduction, high-fidelity sound, and surround sound mixing, reproduction, and projection all became industry standards. Gianluca Sergi termed the Dolby company's intervention in film as the beginning of an era in his book *The Dolby Era* (2004). Although it could be argued that a film technology company cannot instigate a revolutionary era, this technology emerged due to a distinct interest in spatial audio experiences.

This wave of experimentation with spatial audio arrived later in India due to mainstream film conservatism. *1942: A Love Story* (Chopra, 1994) was one of the earliest Indian films to use a Dolby Sound format, becoming a landmark film in surround sound production. The film's sound was mixed at the Dolby SR facility in London. Romantic thriller *Rangeela* ('Colourful', Varma, 1995) followed shortly thereafter and became one of few Indian films

Chattopadhyay, B., *Sound in Indian Film and Audiovisual Media: History, Practices, and Production*. Taylor & Francis Group, 2024
DOI 10.5117/9789463724739_CH10

mixed and released in the Dolby SR format. The company subsequently advanced Dolby Digital technology, now used around the world.

In India innovative and business-savvy filmmakers such as Ram Gopal Varma led the way in incorporating new technologies such as Dolby Digital into their work. Varma utilized Dolby Digital, using its noise reduction and surround sound capabilities, in *Daud* (Run, 1997). Verma incorporated such technologies to produce a cleaner, gentrified and 'corporate' sound design in cinema, in which electronic music with capacity to render low frequency tones played a major role. Verma's film sound work is remarkable in its ability to use sound and music for psychological manoeuvring of his audiences and listeners. His work ushered in a new era of digital film sound exploring the full potentials of Dolby Digital, and later the surround sound technologies.

Meanwhile, David Dhawan's *Judwaa* ('Twins', Dhawan, 1997) used the DTS System, taking advantage of its surround and stereo mixing capacities; DTS (Digital Theatre Systems) was founded in 1993 as a competitor to the Dolby Labs surround sound monopoly.

Within a few years Dolby Digital Surround EX had taken over India's surround sound practice. The technology rapidly led to a more standardized Dolby Digital 5.1 set-up, introducing an additional rear surround sound channel in theatre projection to produce a 5 plus 1 channel set-up. More films embraced the surround revolution, aiming to reach new audiences. In the 2010s, Dolby introduced a new format called Dolby Surround 7.1 for digital cinema, which used eight discrete audio channels (through seven standard speakers and one subwoofer) to establish four surround sound mixing and encoding zones, for reproduction via Dolby decoding in cinema theatres. Indian drama film *Dum Maaro Dum* ('Hit a Puff', Sippy, 2011) was one of the earliest Indian films to mix in Dolby Surround 7.1.

Parallel to these developments, as discussed in the previous chapter, digital sync sound techniques ushered in seismic changes to Indian film and audiovisual media production. Sound could be recorded live on location, eliminating the need for actors to dub their performances in the post-production studio, which had been prevalent since the late 1960s. Among the multitude of early films that used sync sound, I highlight *Lagaan* (Gowariker, 2001) that brought the practice to the fore of mainstream cinema. In the southern film industries, *Hey Ram* (Haasan, 2000) was among other significant films using sync sound, a trend that has now been standardized in Indian cinema. Around the same time powerful filmmakers in the Tamil and Malayalam languages, such as Mani Ratnam, advanced coporate Indian cinema to the next level of successfully marrying commerce, digital technologies, and aesthetics of cinematic narration. Ratnam's tremendous success with his trilogy of films exploring terrorism and politics in India's borderlands, namely Roja ('Rose', 1992), Bombay (1995) and Dil Se… ('From the Heart', 1998) lies in his ability to make use of the digital technologies, e.g., Dolby NR, and his collaboration with revolutionary composers such as A. R. Rahman, who scored most of Ratnam's filmworks in his digital home studio, changing the sounds of Indian film music, via the ingenous digital layering, multitrack recording, and innovative digital processing of traditional instruments to create unheard of sonorities.

Currently, Dolby Atmos technology is replacing other surround sound formats in many cinema theatres as well as for portable devices such as phones and tablets. This development has revolutionary potential. Veteran sound practitioner Anup Dev states in an interview (Kaleebullah, 2016): 'The difference in sound quality between 5.1, 7.1 and Dolby 3D is almost imperceptible. …However, the same cannot be said about Dolby Atmos. The difference in sound quality is huge.'[1] Tamil language film *Sivaji: The Boss* (Shankar, 2007) was one of the earliest Indian films to incorporate Dolby Atmos, but since its arrival, many films in India have been mixed in this format leading to the current trend of films made for the ubiquitous OTT platforms.[2] These surround sound mixes are mostly experienced in home theatres if not in earplugs, headsets, or merely stereo sound systems in computers.

The Auditory Setting in the Surround Environment

Contemporary digital audio technology's capabilities can be broken down into five main aspects: 1) an enormous dynamic range from the softest sound to the loudest; 2) discrete full-frequency channels and their complex routing options; 3) the multi-track digital recorder's ability to separately capture sounds from all corners of a location in synchronization with image-gathering; 4) large storage for keeping sounds, including layers of extra recordings (e.g., environmental sounds, room tone, and other characteristic sounds from location, during and after shooting); and, finally, 5) digital

1 Further reading: interview with Anup Dev in *Between the Headphones: Listening to the Practitioner* (Chattopadhyay, 2021b).

2 OTT or over-the-top is a customized platform for providing television and film content over the Internet on demand to suit the requirements of the individual consumer. OTT platforms started to dominate film distribution, especially in the Covid era.

surround sound design's potential for layering sounds in innovative and creative ways.

How do these new productive capabilities influence the creative practice of constructing film space? And how is this new practice reflected in the audience's experience of sound in contemporary Indian cinema? The spatial aspect of surround sound mixing and reproduction is a key element. As interest in Indian cinema skyrockets in a healthy distribution market, locally and globally, the implementation of spatial audio innovations such as Dolby Atmos needs to be upheld to maintain this resurgent audience. Expressions in reviews of current Indian films such as 'real', 'familiar', 'authentic', 'immersive', and 'great sense of place'[3] indicate a shift within the sound experience of Indian cinema, the proliferation of a new trend, with audiences increasingly relating to sonically believable spatial experiences in the constructed film space and diegetic universe of Indian cinema.

As digital surround design became widely accepted in filmmaking, sound production incorporated newly available digital technologies to its existing set-up. Post-production techniques – editing, designing, and mixing in the studio – became faster. The reproduction of sound in theatres and multiplexes moved toward Dolby 7.1, Auro 3D, Barco, and Dolby Atmos systems. Dolby currently faces an opponent in Auro 3D, which entered Indian cinemas with *Vishwaroopam* (Hasan, 2013). Both companies have developed and offer audio technologies that digitalize, split, and route sound into multiple, surround channels and speakers.

Concerning the spatiality of surround sound, Mark Kerins states that the digital sound system 'is engineered to model a "true" 360-degree multichannel environment where the focal point of the soundscape can be anywhere in the theatre' (2006, p. 43). 'True' in this context underlines the 'lifelikeness' (Rogers, 2013, p. 56) of the auditory setting with a narrative strategy of surround sound organization and spatial rendering to 'construct for us a sense of the material world which the characters inhabit' (Fischer, 1985, p. 239). For Kerins, this practice engenders 'expansion of the cinematic soundfield beyond the screen' (2006, p. 43). He adds:

> To some degree this represents a simple acceleration of established narrative strategies – filmmakers have long relied on ambient sound in the 'surrounds' to set up diegetic spaces, and this trend has certainly continued with movies employing DSS. The difference here is that DSS has encouraged the construction of complex multichannel sound mixes,

3 I'm referring to IMDb user reviews of contemporary Indian films.

> where the different sounds in each speaker channel together create a seemingly realistic and complete aural environment in a way difficult (if not impossible) with monophonic or Dolby stereo sound. (2006, p.44)

How do audiences relate to this 'complete' sound environment and orientate themselves within cinematic, audiovisual, or medial experiences? As Kerins shows, the construction of complex multichannel sound mixes might contribute to an expansion of established practice that 'centres on a strategy of immersion in the filmic environment – audiences are, visually and aurally, literally placed in the middle of the action...in which the narrative processes of cinema...communicate complex perspectives, and dependence on a complex interplay between sound and image to orient audiences' (2006, p. 44). In this surround sound environment, audiences orient themselves through processes of spatial perception and cognition. They feel 'being there', by means of careful navigation through sense modalities, including hearing (Mohler, Luca, and Bulthoff, 2013, p. 90). I have discussed these issues in detail in *The Auditory Setting* (2021a) and elaborated on this sensation of 'being there' in an article on digital film sound (2016).

In many post-2000 Indian films made with Digital Sound Systems, which maximize surround sound design capacities, spatial arrangements of sounds trigger cognitive associations facilitated by the creative and inventive strategies of sync recording and surround design. *Delhi-6* (Mehra, 2009), *Kaminey* ('Scoundrels', Bhardwaj, 2009), *Love Sex aur Dhokha* ('Love, Sex, and Betrayal', Banerjee, 2010), *Dhobi Ghat* ('Mumbai Diaries', Rao, 2011), *Gangs of Wasseypur* (Kashyap, 2012), *Kahani* ('The Story', Ghosh, 2012), *Shanghai* (Banerjee, 2012), *Madras Cafe* (Sircar, 2013), *Highway* (Ali, 2014), *Haider* (Bhardwaj, 2014), and *Dedh Ishqiya* ('One and a Half Passionate', Chaubey, 2014), for example, possess a spatially elaborate and realistic slice-of-life quality. The multi-layered and richly evocative audio information in these films play out a spatial topography of the locations where they were shot in the minds of their audiences, creating a sonic association with the sites in the diegetic story-worlds.

Referring to believability and cognition in cinema, David Bordwell describes the audience as an 'active information seeker' (2009, p. 360) that extracts information from the phenomena of the natural environment they encounter (2009, p. 363). In contemporary Indian films made with surround design, the use of sound brings a wider diffusion of auditory information into the film spaces, adding depth, texture, auditory details, and perspective, so the audience can develop a spatially oriented, enveloping, and expanded sound environment beyond the screen. These possibilities have motivated film scholar Ranjani Mazumdar to claim that in contemporary Indian films even 'the city's wastelands saturate the mise-en-scène' (2009, p. 238). The practice of surround sound in Indian cinema manifests in paying due attention to the spatial dynamics and sonic realities of India's urban and suburban, as well as its rural sites and landscapes.

Atmos(p)here

Many of the films mentioned above explore Dolby Digital and other conventional surround sound formats (Dolby 5.1, Dolby 7.1, and DTS) while more recent films are made with Dolby Atmos and other expanded surround sound formats such as Auro 3D and Barco. Dolby Atmos is essentially an advanced technique in surround sound design to organize extra layers of sounds over regular Dolby Digital 5.1 or 7.1 mixes, which are made using a traditional channel-based method. The layered mix largely consists of atmospheric sounds or actual location-specific ambient sounds, which often used to be removed or subdued as a mere guide track in Indian cinema's dubbing era. Additional sonic elements or an audio object can be positioned on top of this layering according to the corresponding on-screen visuals. Instead of using the speaker channels, Atmos can pinpoint individual speakers through Wave Field Synthesis, making the position of sounds much more precise, moving in-between speaker positions in cinema theatres. The audience can accurately trace sounds within the space such as a moving helicopter in a war scene, an individual bird flying in an outdoor scene, a waterdrop or a mosquito in an indoor scene, or murmurs through the trees in a forest scene. Through their listening capabilities, audience members feel like they are inside a film space, as various spatial and site-specific sounds, including low frequency rumbles and room tones, reflections and resonances of voice and sound effects, fill the space.

If we carefully listen to several of the earliest Indian films in which Dolby Atmos was incorporated – namely *Madras Cafe* (Sircar, 2013), *Bhaag Milkha Bhaag* ('Run, Milkha, Run', Mehra, 2013), *Dhoom 3* ('Blast 3', Acharya, 2013), *Ghanchakkar* ('Crazy', Gupta, 2013), *ABCD: Anybody Can Dance* (D'Souza, 2013), *Nautanki Saala!* ('Dramatic Scamp!', Sippy, 2013), and *Thalaivaa* ('Leader', Vijay, 2013) – we can locate ample examples of how an atmospheric and immersive expanse of sound is imagined and realized via an advanced level of sound design despite economic, methodological, and normative constraints, as well as an initial resistance to experimentation.

Take *Madras Cafe* (Sircar, 2013), for example. The film is set in the turbulent period between the late 1980s and early 1990s, boldly exploring how India intervened in the Sri Lankan civil war and the Indian prime minister's assassination. Its first half is layered and eventful, including complex encounters, diplomatic missions, information leaks, sporadic tension, sabotage, and violent fights between rival parties. This series of events slowly reveals war-torn Jaffna in Sri Lanka, and ongoing ethnic conflict perceived from the perspectives of secret agent Vikram and journalist Jaya – the two unlikely protagonists through whom the audience experiences this setting. It brings to life the Sri Lankan war, highlighting India's ambiguous role, moving sensitively with great restraint in an observer's role, taking no sides. Its second half grows more fraught and fast-paced; the conspiracies are exposed through sound and moving images. *Madras Cafe*'s situated narrative builds up to an agonizing end of loss and trauma, while still finding solace in some exposed truth. The sound was recorded on location with sync sound techniques, capturing naval warships passing through deadly sea, gunfights, and the hollers of people dying, as helicopters float over hills, hauntingly. Mixed using Dolby Atmos, which had just arrived in India at that time, *Madras Cafe*'s enveloping soundscape becomes a shifting site of a psychosomatic experience of war and conflict as military helicopters move from the front of the screen through overhead speakers to the off-screen acoustic space in the rear channels, and gunshots reverberate through this immersive space. The film won two awards for Best Audiography in the 61st National Film Award: Nihar Ranjan Samal won Best Location Sound Recordist and Bishwadeep Chatterjee Best Sound Design.[4] Chatterjee spoke in detail about his approach to mixing in Dolby Atmos for the first time in *Between the Headphones* (Chattopadhyay, 2021b).

As *Madras Cafe* exemplifies, the Dolby Atmos format gives sound designers the ability to move sound as background or foreground objects, on or off-screen, providing a more realistic feel for the audience, surrounded from the front, sides, rear, and overhead. Instead of 5.1, 7.1, or 11.1 channels of surround sound, which Indian audiences were used to from the late 2000s, Dolby Atmos and other 3D audio formats allows filmmakers and sound practitioners access to up to 64 channels. Sophisticated routing algorithms and speaker positioning in the theatre recreate a perceptually realistic experience where sounds seem to move in a very precise manner. Bishwadeep Chatterjee corroborates: 'When we [sound engineers]

4 See the award details: https://web.archive.org/web/20140416181218/http://www.dff.nic.in/List%20of%20Awards.pdf (accessed 4 July 2022).

are panning the sound of a helicopter flying above the audience, ...using Atmos we can pinpoint the source and the sound becomes more realistic as a result. Moreover, we can also pan the sound using overhead speakers' (Kaleebullah, 2016).

While *Madras Cafe* is largely set outdoors, *Ghanchakkar* (Gupta, 2013), in contrast, mostly takes place indoors or inside Mumbai train compartments. In this film, the claustrophobic interior locations vibrate with a constant, low-frequency room tone, emanating from all sides in the Atmos mix, which emulate the protagonist's retrograde amnesia. The audience is aurally trapped as the character gradually collapses alongside the story's sanity. The sillier moments of this comedy thriller take place inside a train compartment: the sounds of a local night train surround the audience, its vibrations, experienced through sonic spatial association, adding to the comic narrative.

Audiographic Realism by Presence and Spatial Association

As seen in the fifth chapter on Satyajit Ray, the sound practice in social realist Indian films provides a detailed depiction of location. It created a realistic and believable context for situated listening through a sense of sonic presence in which the audience found ample information to relate to the places and people that are the subjects of the filmic narrative. I have termed this mode of realism generated by sonic presence in the film space as 'audiographic realism'. Audiographic realism refers to the synchronized use of sound without significant sound manipulation that retains the materiality or the object-hood of the documented sound with perceptual fidelity, leading to a more realistic representation of the original site and situation within a monaural screen-centric use of site-specific sound elements. I mentioned earlier that this is analogous to photographic realism's determination not to affect the appearance of a photographic object (Kania, 2009, p. 240). As discussed, keeping a realist paradigm in cinema was an authorial choice: realist filmmakers used sound as a synchronized element to remain close to reality as a cinematic ethos. As Ray stated, 'the main contribution of sound was an enormous advance towards realism, and a consequent enrichment of the medium as an expression of the ethos' (2011, p. 5). Here 'realism' refers to the tradition of documentary or observational cinema, which represents reality by recording sound that originates 'from within the world of the film' (Kania, 2009, p. 244).

Commenting on the advent of sound recording, previously known as phonography, and its galvanizing sense of realism, Friedrich Kittler notes:

'The phonograph does not hear as do ears that have been trained immediately to filter voices, words, and sounds out of noise; it registers acoustic events as such. Articulateness becomes a second-order exception in a spectrum of noise' (1999, p. 23).

In this statement, the emphasis on 'as such' hints at the revolutionary potential of sound recording to replace static visual scores with recordings of dynamic sounds. Audio recording, particularly early direct sound in cinema and contemporary sync sound in Indian films, carries the same dynamic sense of realism forwards. One may question whether sound recording can be representational, too. In *The Auditory Setting* (2021a), I aimed to show the process of technological mediation embedded in sound recording. What I found in my research is that the embodied sense of presence and situated reality of a live sonic environment often remains elusive in a temporally linear form of sound recording. Likewise, award-winning Indian sound designer Resul Pookutty laments: 'Realism doesn't exist in cinema. It's the real problem of cinema; there is only realistic portrayal of things that make you believe it's real' (Chattopadhyay, 2021b, p. 217). However, when sound recordings are organized in a Atmos or Auro 3D surround sound mix, a sense of spatial presence is better constructed in comparison to a linear monaural or a stereophonic soundtrack. Indian sound re-recordist and mixing specialist Anup Mukherjee concurs: 'In mono we couldn't do anything. All we could do was vary the levels. In case of surround now, it's 5.1, 7.1 and even 9.1, and there is Atmos. So, everything can be kept in space now. It has almost come to reality. Even then, since cinema should be subjective, I will only decide what to make people hear. Just like we hear only what is necessary for us, the rest we generally discard' (Chattopadhyay, 2021b, p. 121). The subjective degree of presence in contemporary Dolby Atmos or other 3D surround sound environments is enhanced on perceptual levels as part of the cognitive universe derived from lived experiences.

If we consider this sense of audiographic realism as a premise to understand the emerging spatial sensibility in digital, multichannel, surround sound production in Indian cinema, the current proliferation of site-specific sounds plays an important role. Sound produced via the intricate digital surround spatialization of these sonic layers[5] creates an enhanced, situated

5 In multichannel experimental soundscape composition and electro-acoustic music creation, both in production and performance, the term 'spatialization' is increasingly used to denote diffusion of sound in space. Its origins can be found in the English translation of the French term 'l'espace', introduced by Henri Lefebvre (1992) with reference to sociocultural perception and cognition of geographical space.

experience, a believable topography relatively close to the lived experience of place in the mind of the audience. It is no surprise that the current breed of Indian films, made with Dolby Digital, DTS, Dolby Atmos, Auro 3D, and Barco, compels the audience to utilize their sensorial, ambient, or environmental faculties of listening instead of looking at the fixed screen. This new realm of sound production supports the emergence of an embodied experience of spatial presence through a cognitive association with places, spaces, and people via auditory associations. This sense of belonging or 'being there' is central to the success of Dolby Atmos. Mixing specialist Anup Dev states: 'Your distribution of sound should be such that one should feel, "I am there." This is very helpful in this Dolby Atmos' (Chattopadhyay, 2021b, p. 79).

Don Ihde articulates embodied experience in this illuminating statement: 'Sound permeates and penetrates my bodily being. ...Its bodily involvement comprises the range from soothing pleasure to the point of insanity in the continuum of possible sound in music and noise. Listening begins by being bodily global in its effects' (Ihde, 2007, p. 45). The pleasure derived from bodily relating to the constructed film and audiovisual media environment is based on how convincing and realistic they sound to the ear. Likewise, the realistic portrayal of fictional places and spaces in Indian digital films with expanded surround sound practice leads to popular appreciation and a sense of euphoria. In user reviews of the popular film *Jab We Met* ('When We Met', Ali, 2007), shot with sync sound techniques and mixed in Dolby Digital surround sound, two viewers underscored the distinctly embodied spatial experience, calling it 'real, natural, and believable':

> The scenes have been mostly shot at outdoor spots like Chandigarh, Kulu, Manali and Shimla, and this entertains us as if we are experiencing a real tour ourselves. Again and again, seeing daily studio scenes made the eyes wounded and fed up the mind [sic].
>
> [T]he execution is so...realistic that no situation in the movie looks out of place. ...absolute[ly] real, natural and believable. (IMDb, 2009)

This euphoria reflects new developments in cinematic experience, acknowledging a renewed sense of realism in how cinematic narration and film spaces are created by the spatial ordering of sounds to 'produce a space for the film to exist in' (Holman, 1997, p. 177). A number of recent Indian independent films such as *Asha Jaoar Majhe* ('Labour of Love', Sengupta, 2014), *Court* (Tamhane, 2014), *Masaan* ('Fly Away Solo', Ghaywan, 2015), and *Killa* ('The Fort', Arun, 2015) do not rely on a musical score, or practically do away with it, using a reduced amount of dialogue (or no dialogue, as with *Asha Jaoar Majhe*) instead. These films are packed with rich layers of actual sounds: street noises, car horns, tram bells, voices of street hawkers, cats meowing and crows cawing, background radio news announcement, and other recognizable actual sounds that are present in the everyday life of Indian cities and rural regions. Due to this careful inclusion and elaborate spatial organization of actual site-specific sounds, these films have a gritty documentary feel, marked by an immersive, immediate realism that contrasts with conventional song-and-dance films from Bollywood. These independent films represent a renewed sense of situatedness in everyday life, meticulously portraying ordinary places and people known through a lived experience of contemporary India. Due to the narrative strategies used, these places and people become characters within the stories, contributing a resounding presence to the film space.

However, even in the conducive creative environment of digital sound production, whether all the subtle aspects of phenomenal worlds from Indian urban and rural locations and environments are faithfully narrated in the produced sonic experience is, of course, a question. On many occasions, noisy parts of sound recordings are controlled and sanitized by editing and noise reduction to provide 'cleaner' sonic environments in cinema.[6] The typically syncretic, chaotic, and inchoate structure of Indian cities – particularly the complex character of everyday urban ambience, including multiple layers of sounds from pre-industrial, industrial, and post-industrial eras, simultaneously active in juxtaposition or contrapuntal relationship with one another (Chattopadhyay, 2014, p. 140) – are not represented. Indian cities are sonically overwhelming and potentially disorienting for the listening subject (Chattopadhyay, 2014, p. 140). Aestheticized accounts of these location are delivered through selected soundmarks instead of faithfully capturing the complete ambience of sites. A 'soundmark', according to Schafer, is a 'community sound, which is unique or possesses qualities which make it specially regarded or noticed by the people of that community' (Schafer, quoted in LaBelle, 2006, p. 216). In the popular Indian road-movie *Highway* (Ali, 2014), the two protagonists (the abducted girl and her fugitive captor) travel through north India in a truck, as outlined in the previous chapter. Every place shown on the screen is established with a soundmark specific to the site often at the expense of many other sounds present to compensate

6 Referring to interviews conducted with Indian sound editors and mixing personnel such as Bobby John and Manas Choudhury for *Between the Headphones: Listening to the Practitioners* (Chattopadhyay, 2021b).

for the noise reduction and editing of multi-track sync sounds in post-production. These 'industrial' norms, rules, and regulations embedded in film industry sound production practices hinder the sound practitioner in applying a more artistic approach that might further enrich the sound experience offered by the film. Despite these apparent drawbacks, pervasive digital technologies such as surround sound endow contemporary Indian films with a scenario for comparatively better practice, incorporating a comprehensive and inclusive approach to sound recording and design methods. Compared to the dubbing era, sound production in the digital era renders narrative spaces far more believable with an enhanced and intensified sense of spatial presence, especially those crafted by surround sound design.

References

Bordwell, D. (2009). "Cognitive Theory". In Livingston, P. & Plantinga, C. (eds), *The Routledge Companion to Philosophy and Film* (pp. 356–365). London: Routledge.

Chattopadhyay, B. (2014c). "Sonic Drifting: Sound, City and Psychogeography". *SoundEffects* 3(3), 138–152.

Chattopadhyay, B. (2016). "Being There: Evocation of the Site in Contemporary Indian Cinema". *Journal of Sonic Studies* 12. Research Catalogue. https://www.researchcatalogue.net/view/286592/286593 (accessed 10 September 2023).

Chattopadhyay, B. (2021a). *The Auditory Setting: Environmental Sounds in Film and Media Arts*. Edinburgh: Edinburgh University Press.

Chattopadhyay, B. (2021b). *Between the Headphones: Listening to the Practitioner* (2021b). Newcastle upon Tyne: Cambridge Scholars Publishing.

Fischer, L. (1985). "Applause: The Visual and Acoustic Landscape". In Weis, E. & Belton, J. (eds), *Film sound: Theory and Practice* (pp. 233–250). New York: Columbia University Press.

Holman, T. (1997). *Sound for Film and Television*. Boston: Focal Press.

Ihde, D. (2007). *Listening and Voice: Phenomenologies of Sound*. New York: Sunny Press.

IMDb (Internet Movie Database) (2009). "Review". www.imdb.com/title/tt1093370/reviews (accessed 1 May 2023).

Kaleebullah, E. (2016, March 22). "Making Bollywood Movies With Dolby Atmos". *Gadgets 360*. www.gadgets360.com/tv/features/making-bollywood-movies-with-dolby-atmos-570913 (accessed 1 May 2023).

Kania, A. (2009). "Realism". In P. Livingston and C. Plantinga (eds), *The Routledge Companion to Philosophy and Film* (pp. 237–247.) London and New York: Routledge.

Kerins, M. (2006). "Narration in the Cinema of Digital Sound". *The Velvet Light Trap* 58 (Fall): 41–54.

Kittler, F. (1999). *Gramophone, Film, Typewriter.* Translated by Geoffrey Winthrop-Young and Michael Wutz. Stanford, CA: Stanford University Press.

LaBelle, B. (2006). *Background Noise: Perspectives on Sound Art.* New York: Bloomsbury Academic.

Lefebvre, H. (1992). *The Production of Space.* Hoboken, New Jersey: Wiley

Mazumdar, R. (2009). "Satya". In Gopalan, L. (ed.), *The Cinema of India* (pp. 236–245). 24 Frames Series. London: Wallflower Press.

Mohler, B. J., Di Luca, M. and Bülthoff, H. H. (2013). 'Multisensory Contributions to Spatial Perception'. In Waller, D. A. and Nadel, L. (eds), *Handbook of Spatial Cognition* (pp. 81–97). Washington, DC: American Psychological Association.

Ray, S. (2011) *Satyajit Ray on Cinema.* Edited by Sandip Ray. New York: Columbia University Press.

Rogers, A. (2013). *Cinematic Appeals: The Experience of New Movie Technologies.* New York: Columbia University Press.

Sergi, G. (2004). *The Dolby Era: Film Sound in Contemporary Hollywood.* Manchester: Manchester University Press.

11. Sound in the Audiovisual Media Arts

Abstract: The eleventh chapter is dedicated to the study of the creative practices of sound in other audiovisual media art productions such as interactive installations, digital media artwork, new media, video and sound art. Several artworks, projects, and initiatives are discussed in this chapter to draw an overview of the nascent but developing field of sound art in the Indian subcontinent due to globalization, planetary mobility, and media community exchanges. Focus is given to young artists who are experimenting with sound in or outside of cinematic contexts, and how these independent practices expand the scope of sound and listening in the contemporary (media) arts fields, challenging their inherent Eurocentrism.

Keywords: audiovisual, media arts, sound art, Indian subcontinent, globalization, decoloniality

Media Arts in the Global Souths

Traditional artistic and cultural sound practices in South Asia have a tenuous relationship with modernist media such as sound recording, photography, film, and radio exported from the colonial West. In the realm of sound practices, for example, recording technology's invasion of South Asia and many parts of the colonized Global Souths[1] had a mutating effect on traditional practices and indigenous technologies (e.g., pre-colonial instruments and tuning systems). From the earliest days of sound recording, modernist technologies shared a fraught relationship with pre-colonial South Asian cultures: local practitioners strongly resisted against their voices and sounds being recorded. Their contention and collective resistance to media imperialism and colonization was first recognized in the early-1970s when so-called developing countries began to criticize the dominance developed countries

1 I add an 's' after the term Global South to underscore the plurivocality of the regions, formerly known as 'third world', sharing a common colonial past and decolonial struggles.

Chattopadhyay, B., *Sound in Indian Film and Audiovisual Media: History, Practices, and Production*. Taylor & Francis Group, 2024
DOI 10.5117/9789463724739_CH11

still held over their media and use of media technologies, a continuation of the control used to quantify and surveil colonial subjects.

And yet, a cultural mélange has also had mutually beneficial aspects: the two spheres have become entangled, coalescing, and adjusting to one another, rather than having a binary and temporally linear relationship (e.g., tradition versus modernity, pre-modern versus modern). If we study the introduction of media technologies to the Global Souths, for example in India, modernist media did have a mutating influence, but colonial interventions also created transformative contexts in which inter and cross-cultural confluences could take place, and hybrid forms of art and aesthetic expressions emerged.

Artists constantly redefine the traditional categories of arts and crafts by incorporating emerging technologies into their artworks. Media art, which flourished in the early twentieth century, often depends on new technological components. The term is specifically used to indicate a certain group of artworks, which are created by recording and modifying sound or visual images. Time-based, audiovisual media artworks change and 'move', in contrast to older art forms that are static. Due to these differences and the underscored definition of the medium (e.g., sound art, film, video art), the medial dispositive is evident. Indeed, the formal and technical aspects of media arts often overpower the content and context of artworks.

This in turn destabilizes traditional artistic practices in South Asia that foreground everyday life, places, and people. Artists are under pressure to find innovative ways of creating a novel medial dispositive or device and more conducive media environments for engaging mass audiences, who are largely uninitiated. Artists are not encouraged to conceive of media artworks where the mediated presence of lived experiences is constructed as a parameter for an artwork's aesthetic appeal. When discussing emerging media arts in India, the departure points are most likely to be notions of situated presence dislocated, mediated, and processed through media technologies.

As a result, media arts, especially sound-based artistic practices, are largely unappreciated in India. Sounds presented in an exhibition, installation, or performative context don't engage the general Indian public; audiences cannot yet imagine artistic forms and aesthetic expressions in which sounds from outside traditional music or performing arts contexts are appreciated. For sound-based artistic practices, there is therefore almost no support from the state, cultural organizations and institutions, and gallery owners. Sound-driven media arts also don't sell well and are difficult to insert into India's mainstream art-world and parochial art markets.

It's no wonder that Indian audiovisual media arts (e.g., experimental film and video art, installation, sound art, and digital art) have had a belated emergence and short history. But interest is slowly rising. Technology-based media artists in India are making a concerted effort to be accepted in the West (e.g., European festival circuits, international funding support, and other forms of dissemination in which artists from India or the Global South don't find a central position), but their work often faces tokenism and isn't critically engaged with enough, given the provincial European standards of arts and cultural production.

Due to the spatiotemporally distinct nature of traditional art in this part of the globe (e.g., temporality, spatiality, and subjectivity),[2] the current rise in sound-based media arts in India can be seen as a response to the East-West (or North-South) conflict and confluence that started in the early part of the nineteenth century through artistic exchanges and cultural transmissions, resulting from Europe's colonial pursuits. Critical and reflective analysis, and the historicization of relevant works – including artistic production, which recognizes traditionally oral and temporally nonlinear art in India, *and* incorporates imported media technologies, conforming to culturally imperialist and modernist pressures of the Global North – can be used to trace this friction, conflict, and confluence.

Digital (R)evolution, Globalization, and Sounding (Media) Arts

The so-called digital revolution in the production and proliferation of audiovisual media during the 1990s coincided with emergent neoliberalism's economic pressure on intensified globalization. The ebb and flow of information, knowledge, and audiovisual content found an emancipatory force in the form of the internet, despite the power imbalances of a post-colonial order within this globalization process. The internet also added a major dimension to the reception, distribution, and marketing of new music, sound works, and other audiovisual media. It created new platforms for media access and promotion, which simultaneously raised questions about authorship, copyright, access, equity, and democracy. Such internet exchange platforms include social media networks and forums, free online sound projects and archives, downloadable web content, peer-to-peer networks,

2 These issues are thoroughly discussed in my books: *Sound Practices in the Global South: Co-listening to Resounding Plurilogues* (2022a) and *Sonic Perspectives from the Global South: Connecting Resonances* (working title, forthcoming 2024).

online music streaming and web-shops, websites for record companies and performers, online music journals, online concerts and interviews, and notifications and bulletin boards.

The relatively freer movement of sound-related content represents new ways of connecting audience/consumers with global/local music, sound, and related audiovisual media from performers within and well outside of Western music industries and companies. Open access to sound was a serious commercial hazard for music companies, but it developed new scope for the innovative distribution and reception of sonic artefacts. Most importantly, freer access to sounds from other cultural settings made it possible to imagine exploratory forms of aesthetic expressions such as sound art and digital art. A whole generation of Indian creators, students, and young practitioners and makers were exposed to the sound-based cultural production and artefacts from Europe and America, where 'sound art' emerged from the historical tradition of Western contemporary art and culture. The specific field of sound art moved from Luigi Russolo's noise intoners and subsequent experiments by Dadaists, Surrealists, and the Situationist International to Fluxus happenings and environments, forming a trajectory that leads to what Seth Kim-Cohen terms 'the conceptual turn' (2009). The work of John Cage and others, made as sound-in-itself on the one hand and as the sporadic inclusion of sound in gallery or museum-based visual art on the other, contributed to this post-war period of sound as art and a conceptual exercise. The apparent diversity of 'sound art' has led to debate about whether the genre falls within the domain of visual art, experimental music, or new media – or falls between categories. The cultural production and dissemination of sound art slowly increased in the 2000s, which has since garnered attention. But, as artist and writer Brandon Labelle claims, this growing activity has occurred rather 'tentatively and ambivalently' (2006).

These waves of creativity in the West have been matched by ripples in the Global South, too. The first decade of this millennia saw sound-based artistic practices gently emerging in India among film school students, musicians dissatisfied with the status quo of traditional music and sound industries, and artists who needed to explore outside the box. Those who were curious and open to experimentation came across a wide range of artwork on the internet through then new virtual communities such as MySpace and later SoundCloud, in which many new sounds as well as kindred local-global alliances could be formed. These networks helped artists to meet and share sounds and inspire one another.

Audiovisual media was an important basis for building social networks within 'glocalization', a term meaning a local and global convergence beyond geographical limitations. Sonic media as channels of communication on the World Wide Web, messengers carrying knowledge and information beyond immediate horizons as facilitators of a new social infrastructure for human interaction, and hybrid forms of aesthesis played central roles in this glocalizing process. Glocalization also formulated a certain rearrangement of media and maintained specific forms of media such as sound recordings to be dispersed over a global space that related to flows, mobility, and the movement of media constituents. David Morley writes in *Spaces of Identity*:

> Globalisation is about the compression of time and space horizons and the creation of a world of instantaneity and depthless-ness. Global space is a space of flows, an electronic space, a decentred space, a space in which frontiers and boundaries have become permeable. Within this global arena, economics and cultures are thrown into intense and immediate contact with each other. (1995)

In this global 'space of flows', the creative exposure to media and sonic experiments elsewhere became a vehicle for local audiovisual experiences. Indian artists, filmmakers, and musicians started to find validation for their own artistic imagination around sound as a creative and exploratory media and phenomenon, which previously lacked context. Various forms and formats of sound works such as noise music, audiovisual performances, sound installation, and field recordings appeared on local scenes. Although there were almost no venues or institutional support in the first decades of the new millennium for this activity, artists reached out to global events, sometimes travelling to festivals in Europe and elsewhere. This resulted in increasing mobility and a disembedding of social acts beyond geographical limits, suggesting a mélange of approaches and methodologies, often subverting local sonic traditions and cultures.

Artists Working with Sound In and Out of India

Insofar as the use of sound as an artistic medium is relatively nascent in India, sound art is gradually becoming an acceptable part of auditory culture. Musicians, composers, media artists, filmmakers, scholars, and listeners are becoming more attentive to an ever-widening variety of sonic experiences, the creative possibilities of sound, spanning electro-acoustic and digital production, via recording, transmission, or playback. Sound art's recent recognition and appreciation started when media scholars active in the arts and cultural studies looked at sound as a marker of cultural difference and convergence; due to the relatively brief history of sound art in India, the number of works is minimal but gradually evolving.

With little financial support, sound artists struggled to make their voices heard in India's contemporary art scene until recently when a few stand out events supporting sound-based artistic practices transpired. Serendipity Arts Festival, for example, hosted a seminal group exhibition of sound works in its 2018 edition in Goa; my installation work *The Unspoken, The Ineffable* (2018) was included along with an array of work by other emerging Indian artists.[3] Khoj and Sound Reasons – both based in Delhi – also support sonic artists.[4]

In 2021 I curated an anthology of sound works from India.[5] The album, which presents a compilation of diverse exploratory music and sound pieces by prominent artists and composers, breaks away from tradition (i.e., Indian classical music, conventional tuning systems, and folk melodies, as well as songs and performing arts). However, the pieces also draw inspiration from this auditory heritage, using existing motifs, sonorities, and textures. The often fragile yet complex relationship between sonic traditions and medial modernity in India is the crucial point of entry to this body of work, which attempts to bridge the two extremes.

Whether the work is deemed 'sound art' or 'experimental' is open for debate, however. These artists perceive sound art as a Western taxonomical construct; Indian musicians and sound practitioners have long experimented, albeit with reservations, to keep their traditions alive and transformative. This compilation newly locates the tension between historical trajectories of sound making and thinking, and the intervention of modernist technologies like recording and mixing that the artists make use of through critical negotiations. It uncovers several questions around de/coloniality and its shadows on emerging sonic practices in South Asia.

The anthology contributes to *Connecting Resonances*, my ongoing research that encourages critical listening to sound practices and auditory cultures in the Global South with an interest in sonic confluences and decoloniality. The works I curated in this release reconsider and rethink durationality (among other issues) in relation to the ephemeral quality and improvization flexibility that Indian sound worlds embody. Invited artists entered a

3 More information on this first-of-its-kind group exhibition: www.serendipityartsfestival.com/archives-2018/event/sounds-in-my-head/ (accessed 1 March 2023).

4 See: https://khojstudios.org/event/word-sound-power/ and http://soundreasons.in/ (accessed 1 March 2023).

5 Bandcamp page of the release: https://unexplainedsoundsgroup.bandcamp.com/album/anthology-of-exploratory-music-from-india (accessed 1 March 2023).

discursive space open to sound practices on the right side of history, where positionality and self-determination enable distinctive Indian voices. Several prominent Indian artists, who often exhibit or perform their work internationally, are giving this history a definite trajectory using diverse approaches rooted in situated practices with strong aspirations for a global reach.

Bidisha Das is a sound-based media artist exploring natural phenomena and the acoustic universe. She works with various self-made instruments and sonification techniques to make environmental phenomena audible. Her work is inspired by the cosmos and bioacoustics. She explores the compound abstraction of soundscapes through several improvized sonic space-time ventures and DIY instruments. Bidisha often captures unheard sounds in nature and focuses on the intangible aspects of everyday experience comprehended through the process of making artwork. During her formative years, she spent a lot of time in wilderness areas, both in India and elsewhere, including the Amazon rainforest. Her work reflects the variety of human and non-human lifeforms that co-exist harmoniously. She has shown her artwork in exhibitions, festivals, and residencies internationally, as well as taking part in the Caltech and NASA space mission design project. While she finishes her studies at KHM (Academy of Media Arts) Cologne, Bidisha often combines art and science to build stories from her exploration with sound deeply grounded in earthy realities.

Surabhi Saraf is an Indian sound-based media artist. Her practice explores complex, performative, and poetic relationships with technology through multimedia works that incorporate sound and video installations, expanded sculptures, performances, and sound compositions. Surabhi was the recipient of the the Artist + Process + Ideas Residency at Mills College Art Museum in 2016, Fleishhacker Foundation's Eureka Fellowship Award in 2015, and the Djerassi Resident Artist Award in 2012. She has had solo shows at Galerie Mirchandani + Steinruecke, Mumbai, India, and Hosfelt Gallery, San Francisco, USA. She has performed at the Thessaloniki Contemporary Art Biennial, San Francisco Electronic Music Festival, NETMAGE 10 International Live Media Festival, Bologna, Italy, and Soundwave Biennial Festival ((5)), Yerba Buena Centre for the Arts and the Asian Art Museum, San Francisco, USA. Her videos have been shown at Times Square, New York, USA, the Blanton Museum, Austin, USA, the Hunter Museum of American Art Chattanooga, USA, and the Museum of Contemporary Art, Vojvodina, Serbia. Her work has been featured in The New York Times, Time Out, Sydney and Mumbai, the San Francisco Bay Guardian, Blouin Art Info, Art Practical, and KQED Arts. Surabhi graduated from the School of the Art Institute of Chicago in

2009 with an MFA in Art and Technology, and recently contributed to *The Listening Biennial*.[6]

Ish S (a.k.a. Ish Shehrawat) is a composer, sound artist, performer, musician, and curator based in New Delhi, India. He works with sound installation and performance of electro-acoustic and improvized music. Trained as a classical guitarist, he has produced sound art installations, albums, and composed music for independent short films, plays, performances, video and media art projects, and contemporary dance recitals in India and abroad. Recently, he presented a 12-channel sound installation at the Yokohama Triennale 2020, Japan. He also curates the Sound Reasons Festival for sound art and experimental electronic music, which has taken place in India and the South Asia region since 2012.[7] In 2018 he received a grant from the India Foundation for the Arts under its Arts Practice Programme to develop a project on the relationship between geometric spatiality in Indian classical music and multichannel sound diffusion.

Nandita Kumar is a new media artist who works at the intersection of art, science, technology, and community to create interactive installations often incorporating sound. She explores the elemental process through which human beings construct meaning from their experiences, by creating sensory narratives through sound, video/animation and performance, smartphone apps, customized motherboards, and solar/microwave sensors, among others. Through her installations, interactive sculptures, paintings, and animations, which seamlessly integrate new media and materiality, Kumar reflects on the striking contradictions within the industrial and natural landscape and environments. She holds a joint B.A. from MS University, Baroda, India, and Elam School of the Arts, Auckland University, New Zealand, and has completed her M.A. in Experimental Animation at California Institute of the Arts, Los Angeles, USA.

Farah Mulla is a multimedia artist based in Mumbai, India. She combines her background in science with arts practice to explore the perception of sound and its effects on human neurology and subjectivity. Often investigating different media, she uses the human voice, field recordings, and other sound elements to explore aspects of human listening experience and the invisible agency of sound via multiple modes of perception. Her research includes experiments with sensory overlaps and materiality through different texts, sounds, and circuits. Having always been curious about sensation and perception processes, her work also incorporates sound brain computer

6 The Biennial website hosts further information: https://listeningbiennial.net/about

7 More information can be found on: http://soundreasons.in (accessed 1 March 2023).

interfacing such as a small adaptive device that detects obstacles in its path and sends a tactile buzz to warn the wearer of the device as mode to develop works of performative installation.

Hemant Sreekumar is a sound and media artist. He got involved with noise music when he started exploring television and radio static in his formative years. He studied art history at the MS University Baroda, India, before becoming a researcher at Compart, Bremen, Germany, documenting early computer generated and algorithmic art. He then began working as a Data Visualization Consultant for W+K. Currently, he uses coded computational processes to create sonic situations. His sound works deal with notions of decay, data sonification, and stochastic emergences.

An Auto-Ethnographic Intervention

In 2005, while still a film sound student at SRFTI, India's national film school, I worked on my first major sound-based art project. I began working independently with sound outside of the film sound canon because I was disenchanted with the way sound production happened then, particularly inside film and music recording studios. I wanted to take sonic technological devices such as microphones and recorders outdoors, on location, into the social milieu and lived environments to engage with the devices and mediation processes personally, from a subjective position of intervention. In the dominant filmmaking practices of India, I felt this artistic freedom was lacking. I believed a more concerned and nuanced approach to working autonomously with sound was needed outside of screen-centric filmmaking. The methodologies I embraced were not very rooted in the vernacular cultures of India but primarily borrowed from the then emerging European sound art scene (e.g., field recording and soundscape compositions). I was mostly exposed to these practices and aesthetic expressions through MySpace, SoundCloud, and other social networks dedicated to sonic and musical experiments, when browsing the internet. I was also influenced by what was termed 'new music' or contemporary classical music in Europe, much of which I could listen to at the local Goethe Institute or Alliance Française libraries. Hardly anyone in India was working with sound outside of film or traditional music canons. I contacted Ish S, who was slowly developing his Sound Reasons festival initiative, in search of local peers and networks.

My 2005 project, *Benaras*, was recorded on an imported MD sound recorder during a journey to the city of Varanasi in northern India. It was produced in my home studio with few resources on a low-cost desktop computer. German label Gruenrekorder later released the work on an audio art compilation CD. The following year I started working on *Landscape in Metamorphoses*, a project set in the undeveloped areas of Santhal Parganas in Eastern India, where I grew up – an area that is endangered by industrialization. Gruenrekorder also released this work on a limited-edition CD.[8]

From these early works, I have developed an auto-ethnographic approach to sound practice, one in which a subjective position of experience drives the work instead of a motivation to observe, measure, and document from the outside. As my body of work has grown over the last fifteen years, I have self-reflexively located a developing interest in situated knowledge, traditional, and ritual practice as an Indian artist operating on a global platform. When I engage with other Indian artists working with sound today, I see a similar thread in their work. They also negotiate the use of imported sound recording and production devices through personalization, hacking and DIY interventions, expressing subjective realities in sound through a situated and embodied perspective. Critically engaging with my own work with its auto-ethnographic approach is useful to shed light on emergent methodologies and the trials and predilections of emerging artists.

Four Corners (2008) was my first exhibition at the then Religare Arts gallery in Delhi, and *Eye Contact with the City* (2010) was my first major solo sound installation. The latter installation was realized in 4-channel sound and 2-channel video projection, produced during the BAR1 residency in Bangalore supported by the India Foundation for the Arts. The work was nominated at PRIX Ars Electronica 2011[9] and later released on CD by Gruenrekorder as a sound composition, *Elegy for Bangalore* (2013), which I have performed in various contexts. The work has been reviewed as being abstract with 'immersive, meditatively palpable qualities' (Jakimowicz, 2010), where the urban site of Bangalore has been taken as a point of departure to enter into 'the immortal capital city of a parallel realm' (Chuter, 2013). The primary materials used in the work were shot video footage, retrieved sounds from archival reel-to-reel tapes found at the city's flea markets, and field recordings made at urban development sites in Bangalore, including metro rail construction sites. To another reviewer, 'this work offers a cinematic walk through the different alleys, neighbourhoods, voices and lives of Bangalore, ...and offers a beautiful reflection on a city that can only be imagined' (Papadomanolaki, 2013). Bangalore, dealing with problems of

8 See: www.gruenrekorder.de/?page_id=301 (accessed 1 March 2023).

9 More information: https://budhaditya.org/projects/eye-contact-with-the-city/ (accessed 21 July 2023).

urbanization, appears to an audience of the exhibited or performed sound work as a parallel site, an imaginary city rather than one that has been documented. The repository of field recordings and other audio material, gathered during my inter-subjective interaction with Bangalore during the artist residency, eventually took the form of this elegiac composition during a subsequent residency at the School of Music, Bangor University, Wales, in 2011. The resulting composition, *Elegy for Bangalore* (2013), made for stereo and multi-channel sound, incorporates situated sonic experiences in an Indian city from the inter-subjective position of an insider.

Decomposing Landscape (2015) is an audiovisual media artwork that offers an in-depth listening to the rapid transfiguration of indigenous habitats and livelihoods along with rural landscapes in India undergoing environmental decay and ecological breakdown. Using on-site field recordings mediated through multichannel diffusion, the work presents a multilayered narrative on deep climate destruction in India, engaging its audience in grounded realities through affective means. The work was developed from a meticulous collection of locative sounds recorded at an SEZ (Special Economic Zone) in the Eastern part of India during extensive fieldwork over several years. The collection forms a digital archive that was instrumental in realizing the work, which was composed, mixed, and produced at ICST, Zurich University of the Arts, Switzerland, during an artist residency in 2014. The work was released in 2015 by British label Touch as both Binaural and Ambisonics mixes.

Dhvāni (2021) is a series of responsive, self-regulating, and autonomous sound installations driven by Artificial Intelligence and Machine Learning, advocating for values associated with interconnectivity, codependence, network, and community from a decolonial approach. *Dhvāni* incorporates the sounds of ritual and traditional practices from South Asia (e.g., temple bells, Buddhist Gongs, wind chimes, and Ghungroo) to bring new perspectives of inter-subjective interaction and autonomous agency in contemporary media arts. The project emerges from a trajectory of artistic research re-listening to and re-telling South Asian listening cultures in a contemporary moment of societal crisis, informing the AI-driven surveillance and controlled societies of today about the values of interconnectivity, community, and reciprocal ways of life, often found in ritual practices of the Global Souths. The work envisions geological equity, rendering the linearity of a Western-modernity-dominated sense of temporality into a cyclical pattern, refocusing on memory, rituals, locative traditions, and indigenous cultural practices. This temporal mélange may help find answers to the crises of today such as climate breakdown and global inequality. When exhibited, *Dhvāni* creates fertile, evolving, and autonomous situations, which are relational, performative, and radically participatory. The audience is an inclusive element in the work, encouraging a reciprocal approach within a shared artistic experience through a network of traditional objects. *Dhvāni* is a neural-like network of temple bells of various sizes and tuning, among other pre-modern ritual objects, that plays a collective sounding when the audience's presence activates ambient sound sensors and, in turn, a number of AI-driven robotic arms.

In all three of the sound artworks described above, tension between the site-specific evidence of everyday life and compositional abstraction engages the audience as a sensate listener and participant. Take for example *Elegy for Bangalore* (2013), whose repository of field recordings and other audio material took the form of an elegiac composition infused with traces of situated listening, gathered through sonic drifting, reflecting the perceived longing for the past embedded in the rapid modernization of India. My works aim to create a conceptual, practical, and methodological premise for in-depth listening to the passage of time. It offers a psychogeographic reflection on emergent urban public sites in India. Chaotic, noisy, and hybridized sonic environments are given a transcendental and cathartic mode of listening. Emphasizing inter-subjective, participatory, and adaptive auditory perception, my works suggest an auto-ethnographic method of listening and sounding to share a grounded and context-aware sonic portrayal of inter-subjective interactions, often retelling rituals in the present moment.

Sound in Games, VR, and Emerging Media Arts

Let's now take a glimpse at creative practices with sound in commercially viable audiovisual media. If we compare digital surround sound in cinema with the sound used in interactive media or gaming industries in India, we may find converging methodologies despite industry specifics. In contemporary media environments, cinema is no longer culturally dominant (Shaviro, 2010). Several other audiovisual forms such as computer games and VR/AR widely practice sound production and consumption. The proliferation of these media practices, emerging in parallel to cinema's more established arena, have changed spatial experiences by altering our relationship with sound. The audience/player/user of games no longer merely navigates and negotiates the mediated space but becomes an active participant, employing multimodal senses to construct the conditions of 'being there'. The sensory stimuli generated via interaction with these media forms and their usage may lead to the imaginary construction of sonic space. I term this space the 'pseudo-site', indicating a difference between both the fictitious/illusory world of the mediated gaming environment and augmented audiovisual VR/AR environments, and independent films made with a realistic, social issues-based story.

In considering the current state of sound production in India, we may initiate a comparative study between those digital (surround) sounds employed in gaming or interactive media platforms and the surround sound used in filmmaking. Here I make a note of the word 'surround', relating it to multi-channel sounds rather than its wide commercial usage (e.g., Dolby or DTS surround mixing and projection). According to Vishal Gondal, CEO of Indiagames Ltd.:

> India behaves comparatively differently than our neighbours like China and Korea when it comes to games. The biggest gaming platform in India is not the PC or the Console, it's the mobile phone. India now has about 700 million mobile users, and this number is growing steadily, and a bulk of these users are actually using their mobile phones to play games. So yes, the biggest platform is mobile. At the same time we are also seeing the growth in the broadband sites, and more and more people are buying the cheaper netbooks and the low-end PCs, which they are connecting to the Internet. And what is really gaining in a big way is Facebook. Facebook gaming is actually becoming very popular...Thirdly we are also seeing a lot of excitement around the set-top-box gaming. People who have television with boxes in their homes are using their remote controls to play games. (Iniarra, 2010)[10]

On the one hand, Gondal's detailed observation shows that digital surround sound is difficult to be incorporated and experienced in India's primarily mobile, tablet, and TV set-top-box-based online gaming market. On the other hand, research shows that 'surround sound players will experience more presence (engagement, spatial presence, social richness and perceptual realism) than two-channel sound players' (Skalski and Whitbred, 2010, n.p.). 'Presence' here relates to the 'perceptual illusion of non-mediation' (Lee, 2004, pp. 27–50) by human-made technology and describes how the audience/player/user gets past the technological tropes and mediation of reality to establish a sense of 'being there', which is often constructed by

10 Interview with Vishal Gondal, CEO of Indiagames Ltd, the largest gaming company in India. The interview was recorded by Marcelo Iniarra at TED Global in Oxford UK, July 2010; available on YouTube, www.youtube.com/watch?v=wSZniXx-3bc (accessed 1 March 2023).

audience participation rather than being readymade. Surround sound in video games therefore needs to affect spatial presence by representing more vivid and elaborate information about sound fields that place the player/user more in the action of a game. Surround sound headsets are popular with gamers because they reflect the 'first-person' genre of games or FPS, where players enjoy a point-of-view and point-of-audition perspective. According to many gamers, in surround sound hearing those footsteps is a big advantage. Media researchers including Bracken and Skalski have conducted numerous experimental studies and concluded that using surround sound in games can potentially contribute to spatial presence experienced during play. More recently in India, younger filmmakers like Sairam Sagiraju and Lijo Jose Pellissery have been exploring the VR platform and developing nonlinear content for VR/AR/XR. Filmmaker Anand Gandhi and others founded the Memesys Culture Lab in Goa, which launched six VR documentaries on ElseVR, a VR journalism platform co-founded by Zain Memon. Gandhi and his team commissioned documentary filmmakers to make issue-based nonfiction films on VR. Matterden, Enlighten, have opened one of the first VR Cinemas in Bandra, Mumbai. Kerala-based Progopanda Productions have been working on a series of immersive VR experiences based on Indian religious texts.

As Vishal Gondal informs us, most Indian interactive games are made for mobile phones and online platforms like Facebook and come with mostly stereophonic sounds relying primarily on digital sound effects and layers of music. Let us examine some popular representative games such as *Bombay Taxi*, *Jhansi Ki Rani*, and *Diwali*. These games are played with sound effects and accompanying music only. The player/user does not have any possibility of engaging with authentic sounds or site ambience due to the absence of any such layers in the sound experience. Instead, the sense of 'being there', through the perceptual realism of a 'pseudo-site', is constructed via social presence (Skalski and Whitbred, 2010).

There are many examples where games overlap popular films in India to share audience attention. The film *Ra.One* (Sinha, 2011) has been turned into a popular game *Ra.One – The Game* by Sony Computer Entertainment Europe. *Krrish* (Roshan, 2006) has also been made into a game of the same name. Surround sound has been employed to a certain degree in these games: sound effects are provided with the movement of actors and, in some cases, manipulated voices (as in the upcoming 3D version of the film *Krrish 3* (Roshan, 2013) and its corresponding game that has already been released). However, experiencing the game leads to the conclusion that the 'effects-music-manipulated voice' sound strategies in these games operate in the mind of the player. A sense of 'being there' is triggered by emotional and affective engagement with a 'pseudo-site' rather than the use of sensorial and cognitive association with a real site. Nevertheless, recent game-like films *Sivaji 3D* (Shankar, 2012) and *Vishwaroopam* (Hasan, 2013), where action speaks louder than words, sparse use of site-specific sounds in Auro 3D and Dolby Atmos sonic spaces, respectively, create perceptual realism. It is a 'sensation of reality' (Skalski and Whitbred, 2010), of 'being sited' as discussed in previous chapters. In India, mainstream films and popular games or emerging VR works complement each other, share many traits, and are in synergy due to a general escapist tendency, hiding from the everyday reality that their sound-worlds are built from.

References

Chattopadhyay, B. (2022a). *Sound Practices in the Global South: Co-listening to Resounding Plurilogues*. London: Palgrave Macmillan.

Chattopadhyay, B. (2024). *Sonic Perspectives from the Global South: Connecting Resonances* (working title, forthcoming).

Chuter, J. (2013, April 19). "Review: Budhaditya Chattopadhyay – Elegy for Bangalore". *ATTN:Magazine*. www.attnmagazine.co.uk/review-budhaditya-chattopadhyay-elegy-for-bangalore/ (accessed 3 July 2022).

Iniarra, M. (2010, July). "Interview with Vishal Gondal, CEO of Indiagames Ltd." *TED Global, Oxford UK*.

Kim-Cohen, S. (2009). *In the Blink of an Ear*. New York: Bloomsbury.

LaBelle, B. (2006). *Background Noise: Perspectives on Sound Art*. New York: Bloomsbury Academic.

Lee, K. M. (2004). "Presence, Explicated". *Communication Theory* 14(1), 27–50.

Morley, D. and Robins, K. (1995). *Spaces of Identity: Global Media, Electronic Landscapes and Cultural Boundaries*. London: Routledge.

Papadomanolaki, M. (2013, October 13). "Review of *Elegy for Bangalore*, Gruenrekorder (Budhaditya Chattopadhyay, 2013)". *Field Reporter/Sonic Field*. https://sonicfield.org/310/ (accessed 1 March 2023).

Shaviro, S. (2010). *Post Cinematic Affect*. Ropley, UK: John Hunt Publishing.

Skalski, P. & Whitbred, R. (2010). "Image versus Sound: A Comparison of Formal Feature Effects on Presence and Video Game Enjoyment". *PsychNology Journal* 8(1), 67–84.

12. Parallel Sounds, Radical Listening – Part II

Abstract: This chapter starts with a discussion of feminist filmmakers in India and their impactful contributions to the use of sound. It then elaborates on independent films of the digital era, often termed 'Indian indies', following Parallel Cinema's inclination toward audiographic realism, expanding into the use of sync sound and location recordings. Later, the chapter takes a critical overview of the use of sound in documentary, short film, experimental, and arthouse cinema genres throughout India. Examining the experimental processes of sound recording and design that widen audiovisual media's scope reveals modulations of reality that resonate with contemporary India.

Keywords: Indian indies, feminist filmmakers, Cinema of Prayoga, non-fiction sounds

Feminist Filmmakers and their Sound worlds

Indian film industries, both in Mumbai and in the regions, have historically been male dominated, but the remarkable work of feminist filmmakers across India challenges the patriarchal genesis of Indian cinema, bringing with them cinematic idioms that push against the normative margins. In this small subchapter, only a very limited number of directors can be partially discussed, but hopefully this will encourage a future film (sound) scholar to expand this analysis.

One can start with Arundhati Devi, who made films like *Chhuti* ('A Vacation', 1967), *Megh o Roudra* ('Clouds and Sunshine', 1969), *Padi Pishir Barmi Baksha* ('Burmese Box of Aunt Padi', 1972), *Deepar Prem* ('The Love of Deepa', 1983), and *Gokul* (1985). Her portrayal of familial voices, often marginalized in society, and their greater aspirations in mundane situations receive

Chattopadhyay, B., *Sound in Indian Film and Audiovisual Media: History, Practices, and Production*. Taylor & Francis Group, 2024
DOI 10.5117/9789463724739_CH12

significant perspective through sensitive listening. The balanced handling of sound contributes to the acoustic ecology of these filmic universes.

Aparna Sen, director of *36 Chowringhee Lane* (1981), *Paroma* ('The Ultimate Woman', 1984), *Sati* (1989), *Yugant* ('What the Sea Said', 1995), *Paromitar Ek Din* ('One day of Paromita's', 2000), *Mr. and Mrs. Iyer* (2002), and, more recently, *Sonata* (2017) and *Ghawre Bairey Aaj* ('The Home and the World, Today', 2019), has spearheaded a feminist and often intersectional-feminist perspective in Indian cinema. Her films, while incorporating conventional production technologies and aesthetics such as dubbing, ADR, Foley, and an expressive use of music, address cultural taboos, societal trauma, and the position of women. Her work also advocates for a collaborative spirit, sense of community, and convivial relationship between fellow filmmakers and practitioners. Through an anecdote from sound technician Hitendra Ghosh, we get a glimpse of the conviviality that Sen tends to embrace in and outside the studio:

> I remember when I was doing *36 Chowringhee Lane* with Aparna Sen, I told her, 'See, I am showing it to you right now. If you want to say anything, say now because now I am going into the optical. So, I will not be able to do any changes.' So I was doing it and the moment the optical was over, Aparna said, 'Hitu, you take out the music from that portion.' I said, 'Aparna, what are you doing? I have already done it! That's why I asked you before!' She said, 'Now I feel that.' I said, 'You can't do it now. I can't do it.' I must have shouted at her, that kind of a situation. She started crying! Then I said that let's put another negative, I can't do anything else. Since she was crying, I had to do it. ...So I did it. The next day I was doing another reel. But she was in a good mood or something that I don't know. She again said, 'Hitu I want this dialogue to be removed.' She was expecting me to shout. She was pulling my leg actually the next day! (Chattopadhyay, 2021b, p. 144)

Mira Nair's work delves into the South Asian diasporic community, issues of migration, and transnational identity. Keen observations and grounded listening, for example in *Salaam Bombay!* (1988), *Monsoon Wedding* (2001), *The Namesake* (2006), and *The Reluctant Fundamentalist* (2012), facilitate studies of the in-between spaces of pan-Indian identity. She tackles issues that are rarely addressed such as child labour in *Salaam Bombay!* (1988), domestic sexual violence against minors in *Monsoon Wedding* (2001), and the social alienation and struggles of South Asian migrants in America in *The Namesake* (2006).

Kalpana Lajmi's *Rudaali* ('Female Weeper', 1993) incorporates locative sounds and music, especially Bhupen Hazarika's situated compositions, to explore themes of gender inequality, casteism, class division, and poverty in rural Indian societies. These social realities are depicted by exposing the patriarchal and feudal systems and socio-economic marginalization of the poor and downtrodden in the cinematic narration. The focus on the female character and her voicing is precursor to several female-centric films that followed, particularly in the present body of independent Indian films. The delineation of resilience in the female character provides hope for an equitable future.

In more recent films, India's celebrated women directors explore contemporary themes such as the childhood and psychological trauma of growing up girls in India's patriarchal middle-class families and overcoming physical disabilities. In their work sometimes sound becomes a probing tool or a methodology for socio-spatial inquiry. Shonali Bose's *Margarita with a Straw* (2015) and *The Sky Is Pink* (2019) delve into these social and domestic issues with sensitivity, restrain, and very insightful humour. Often the use of locative sounds helps reconfigure the listening positionalities of the charcacters. In *Dear Zindagi* (Shinde, 2016), the auditory setting of the sunny, mellow beaches in Goa becomes a playground for the female protagonist, finding solace and sanity away from her/their hyper-modern world, and discovering solidary relationality.

There are many other Indian directors expanding the oeuvre of feminist cinema, including: Shobhna Samarth, Sai Parānjpye, Sumitra Bhave, to name a few, and, from a younger generation, Deepa Mehta, Meghna Gulzar, Reema Kagti, Zoya Akhtar, Tanuja Chandra, Ashwiny Iyer Tiwari, Manju Borah, Leena Yadav, Kiran Rao, Rituparno Ghosh, and Nandita Das, among others. Each contribute to the canon of Indian cinema with their unique cinematic language. Their work deserves further critical consideration.

Independent Films of the Digital Era

In chapter 9, I discussed how Indian cinema has revived observational techniques since the 1990s' digitalization of film technology. An emergent realization of authentic locations and spatial exploration in films suggests Indian cinema's rediscovery of its realist roots. In the past twenty years or so, film sound in India has experienced a massive shift from analogue to digital. This development has initiated the possibility of translating everyday realities in India through intricate processes of spatial sound use. Location sync sound recording is a direct descendent of this trend. Since the success of *Lagaan* (Gowariker, 2001), which implemented sync recording and Dolby digital technology after much debate in the filmmaking community, sync sound has been accepted as a highly precise, artistically demanding, and skilled recording technique in India. Sound technicians can now record original dialogue performed on location along with sited sound effects and ambient sounds, eliminating the tedious process of dubbing or ADR in post-production.

Equipped with such liberating technologies, various directors want to throw off the shackles of mainstream filmmaking industry, loosely termed *Bollywood*, which squarely focuses on investment and profitmaking as opposed to fine aesthetics.[1] Filmmakers such as Nagesh Kukunoor, Dibakar Banerjee, Anurag Kashyap, Hansal Mehta, Neeraj Ghaywan, Sriram Raghavan, Vishal Bhardwaj, Nandita Das, Kiran Rao, Rajat Kapoor, Anand Gandhi, and Anubhav Sinha lead an independent thread of Indian cinema, widely considered equivalent to the Parallel Cinema movement from the 1970s and 1980s. However, if we study these new independent films vis-à-vis the earlier breed of parallel films, a strong focus on film marketing, which particularly exploits global distribution, and a lack of or rather loose commitment to social and political issues, sets today's filmmaking apart from India's earlier parallel films. These independent films jointly negotiate finances and aesthetics prudently as state support, from the National Film Development Corporation, and other funds dry up. The rise of the OTT platforms such as Netflix and Amazon Prime fuel these aspirations while directly provoking homogenous film productions with little aesthetic value.

Analysing the sound from Indian independent films made in the last decade such as *Love Sex aur Dhokha* (Banerjee, 2010), *Dhobi Ghat* ('The Washerman's Ghat', Rao, 2011), *Shahid* ('Martyr', Mehta, 2013), *Haider* (Bhardwaj, 2014), *Labour of Love* (Sengupta, 2014), *Ship of Theseus* (Gandhi, 2015), *October* (Sircar, 2018), and *Article 15* (Sinha, 2019) reveals that audiences are now exposed to the synchronized and direct recording of locative sound that faithfully translates socio-political situations and site-specific realities into the filmic space. These films utilize a documentary feel to convey information, meaning, and spatial depth in the narration. They have much to say and do so through creative sound practices. Microphones receiving

1 *Bollywood* is a casual and loose name given to the Bombay-based, Hindi-language film industry. Bollywood proper is only a part of Indian cinema, which includes other industries producing films in many other regional languages. This book discusses a plethora of films and audiovisual media works as case studies to counter this awkward term.

the actual sounds of India's urban and rural sites, which themselves have many stories to tell, are at the core of this creative approach. Most public spaces in India are deeply layered with multiple sources of atmospheric sounds coming from their historical past mixed with the present, as shown in my article *Sonic Drifting* (2014c).

This new breed of Indian independent films, known as Indian indies, methodologically tends to distance itself from mainstream Indian cinema, known for its typical narrative tropes of spectacular yet escapist song-and-dance extravaganzas. Recent films such as *Asha Jaoar Majhe* ('Labour of Love', Sengupta, 2014), *Court* (Tamhane, 2014), *Masaan* ('Fly Away Solo', Ghaywan, 2015), and *Killa* (Arun, 2015) do not rely on music, or practically do away with it, using a reduced amount of dialogue (or no dialogue, as with *Asha Jaoar Majhe*) in the narration instead. As discussed in chapter 9 on the digital era, these films are packed with rich layers of actual site-specific sounds. Due to this careful inclusion and elaborate spatial organization of actual site-specific sounds, these films are marked by an immersive, immediate realism that stands in strong contrast to popular musicals. These independent films represent a renewed sense of aural situated-ness in everyday public life, comprehensively portraying ordinary places, spaces, situations, and people, including lived environments, socially marginal characters, and ignored or forgotten sites such as small towns and remote villages from contemporary India, with its emerging public urban spaces and transformative rural hinterlands. Due to their narrative strategies of situated listening to the place, these sites and social constellations become another character within the diegetic narrative, contributing a resounding spatial presence. The morbid but emancipatory burning ghats of Benaras in *Masaan* ('Fly Away Solo', Ghaywan, 2015), the cramped city of Dhanbad in the coal mine regions of India in *Gangs of Wasseypur* (Anurag Kashyap, 2012), a northern Indian small town in *Bareilly Ki Barfi* ('Bareilly's Barfi', Ashwiny Iyer Tiwari, 2017) appear palpabale through their sounds.

After OTT took over film distribution and projection across India since the Covid period, and as films are more and more watched and listened to indoors, at home, multiplying their reach and target audiences, a sizeable number of commercial films embrace experimentation, capitalizing on the length and breadth of these new audience bases and their increasing demand for new content. Recent independent film *Blindfold* (Karamen, 2023) is touted as India's 'first' audio cinema with no visuals. In this experimental film, sound has been explored as the main storytelling device. The film revolves around a visually challenged person who is a principal witness in a murder case. As the murder investigation is narrated without moving images, leaving no visual information on the screen, the creative use of sound puts the audience in the ears of the blind protagonist. Beyond obvious gimmicks and market stunts, such films draw the attention of the ordinary movie-goer to the potentials of sound in film and audiovisual media.

Non-Fiction Sounds

Mani Kaul, in conversation with Udayan Vajpeyi, revealed his philosophy on documentary cinema and how he aims to blur the category, boundaries, and strict taxonomies between non-fiction and fiction, through subjective poetic intervention:

> All my life I have tried to find different ways to do away with a linear narrative. This is why I was interested in documentaries. I find the documentary form very interesting, and within it the poetic documentary – the way I made Dhrupad, Siddheshwari, Mati Manas...all of which you might call documentaries, but it would be inaccurate because they should be referred to as poems. In my own way, I have tried to bring poetry, documentary and fiction together. We would call it non-linear narrative. I have a great disdain for the linear narrative that exists, even if it is complex, even if there are many characters, even if it has a large canvas...but the spine is a linear narrative, and if I remove that and the entire thing falls like a pack of cards, then there is no point. I think there is a bug in your mind that demands that it should start from here and finish there. It is the bug of linear narrative. The experience of life is not like this. (Kaul, 1991)

Scholars and practitioners have variously discussed the fragile boundaries between fiction and non-fiction films. The line that distinguishes films with a comprehensive documentary aesthetic, aiming to record reality, from those whose fictional sites promise an escape from reality, from the 'here and now', is often blurry. Indeed, it is often film practitioners themselves who devise work that merges these boundaries. Their films rarely remain as fact-finding and fact-recording but aspire to more. Fact and fiction are blended in inventive ways to convey their worldview. Irrespective of this, however, there is still a taxonomic thrust on defining 'documentary film' via the colonial model of ethnography and othering.

Current Indian documentary films are rich and diverse in practice and handle a wide variety of subjects. Nevertheless, as Giulia Battaglia argues in *Documentary Film in India: An Anthropological History* (2017), Indian documentary filmmaking was born of the colonial model of fieldwork embraced by ethnography and anthropology. B.D. Garga's excellent book *From Raj to Swaraj: The Non-Fiction Film in India* (2007) also delineates how contemporary documentary practice in India follows the work of Hiralal Sen, S.N. Patankar, J.F. Madan, and Harishchandra Bhatwadekar, who pioneered the newsreel form of documentary filmmaking during the colonial era. Indeed, Indian documentary filmmaking can be traced back to the very early days of cinema's arrival in India as an optical media for observing and controlling colonial subjects during the British Raj. *Swadeshi* films, a national form of auto-ethnography, subsequently appropriated this cinematic device. Ashish Rajadhyaksha writes about the historical moment when S.N. Patankar, who was hired to document the 1911 coronation of King George V and Queen Mary in India, turned the camera on home-grown Maratha warrior Shivaji Bhosle (2016, p. 13).

Since then, documentarians and social chroniclers have taken up multiple Indian subjects through disparate approaches to documentary filmmaking, often experimenting with the form, focusing on issues head on with guerilla-style filming. Leading figures such as Anand Patwardhan, Sai Parānjpye, Pankaj Rishi Kumar, Paromita Vohra, Saba Dewan, Rajiv Mehrotra, Shabnam Virmani, and, from a younger generation, Leena Manimekalai, Vinay Shukla, and Payal Kapadia, among others, join a body of work already enriched by India's cinematic luminaries, who also made award-winning documentary films: Satyajit Ray, Mrinal Sen, Ritwik Ghatak, Shyam Benegal, Saeed Akhtar Mirza, Mani Kaul, Kumar Sahani, Buddhadeb Dasgupta, and Goutam Ghose, among many others.

While this subchapter isn't intended as a comprehensive history of the documentary film or documentary sound aesthetics in India – other researchers may later take up a more focused study – I would like to take this opportunity to briefly underline a few conceptual issues about documenting sound. The realist, unmediated aesthetic of documentary practice demands that sound is always recorded in sync. Documentary films used sync sound before it was adopted in other forms of Indian cinema. Audiovisual redundancy (Chion, 1994) is most valid in documentary films' aim to 'capture' reality as it is. For many filmmakers, non-diegetic sound has no role in documentaries. If non-fiction film should document, it makes no sense to add an external sound to a factual representation of the situated image. According to demands of cinéma vérité and direct cinema, only synchronous and diegetic sound is permissible. In 1951 Ray wrote:

> For a popular medium, the best kind of inspiration should derive from life and have its roots in it. No amount of technical polish can make up for artificiality of theme and dishonesty of treatment. The Indian filmmaker must turn to life, to reality. (1976, p. 127)

This statement is the cornerstone for a typical documentary filmmaker. However, the process of artistic transformation in documentary films, from location recordings at a site, or in a situation under critical observation, to producing an augmented environment, needs to formulate narrative diegesis between the abstraction and representation of the sited source. The degree of artistic transformation depends on the amount of spatial information retained from unedited location recordings to studio processing (i.e., editing and sound mixing). The resulting tension may help retain the attention of the film's audience, who might not otherwise lend their ears to a purely documentary representation of the site in raw recordings. I would like to underscore this tension between offering an evidential account through the use of unedited and relatively unprocessed recordings, and an abstraction of sounds, brought in through deliberate artistic interventions and transformation. As a media artist and researcher, I consider this inherent tension makes such works more engaging.

This approach brings us to the concept of 'Cinema of Prayoga', a term coined by film historian and curator Amrit Gangar in 2006 to denote an underexplored, non-mainstream but consistent thread of Indian filmmaking, which challenges the traditional art-commerce binary and complicates western nomenclatures such as experimental or avant-garde films. Cinema of Prayoga advocates a purer, situated practice using applied forms of aesthetics drawn from vernacular cultures and traditional knowledge systems. In associated works filmmakers often break down the synchronous mode of audio and visual relationships, and open the dynamics between them, to make a subjective and poetic intervention. This position of the subjective echoes Kaul's call for poetic cinema, making a strong point to dilute the taxonomies of fiction and non-fiction film.

As I have been attempting to show throughout this book, Indian cinema has adopted two fundamentally different approaches to sound production: namely, synchronized sound or sync sound, and post-synchronization (i.e., dubbing and post-production). In the scope of a synchronized sound practice, clear and detailed sound perspectives are recorded 'on location', from sites that are depicted within the diegetic universe of the cinematic narrative. The 'site' comes alive in the visceral sound world created carefully and confidently from location-based sync sound recording. However, while listening to films made with sync sound, one may wonder if a parallel narrative can be derived from sounds related to what we obviously see, transcending the depicted reality on the screen. Perhaps the narrative devices of sync sound are bound to produce an embodied reality invariably connected to the site-specificity of the image. In pure sync sound production, sound appears to lack the ability to develop an autonomous narration since the recorded sound and the image of the site are mimetically joined together.

Post-synchronization practices like dubbing, on the other hand, open a plethora of possibilities, namely through the deliberate disjuncture between the suggested (but not so credibly depicted) sites in the narration and visual and sonic realities. The listener is given the chance to rearrange multiple potential meanings from an aural cue, since the juxtaposition between sound and image is left loose and flexible in dubbing-based film productions. Often such autonomy in the hand of the listening subject renders the sound world self-indulgent, and a commitment to reality might be lacking in such films. I would argue that the lacunas left in clear meaning-making could be productive and fruitful if a potential listener approaches such films with innocence. Furthermore, such an approach tends to alter cinema's modernist 'dispositive' towards an improvizational mode of listening and looking, wherein the subjectivity of the listener and onlooker can easily intervene.

To unpack the term 'cinematic dispositive', we need to go back to the formulation of the term 'dispositif' by Michel Foucault (1977). In using this term, he describes the mechanisms put in place by an institutional mode of production in any form of knowledge structure that affects the social body. When we use the term in a cinematic context, the various forms of devices, machineries, tools, and apparatus of filmmaking are underscored. For example, both 'dubbing' and 'sync sound' are current cinematic devices in India that deeply affect and influence the dominant working mechanisms of directors, technicians, producers, and funding bodies. Twenty-five years ago, it would have been unimaginable to consider sync sound as a dominant mode of production for Indian films made within set industry norms. When a social body such as the technician's guild approves and promotes a method, it becomes a mechanism that should be followed by all. Dubbing enjoyed its monopoly between the 1960s and 2000s, after which the digital revolution in filmmaking paved the way for sound becoming more 'sitely' and elaborately recorded on location with multi-track capabilities. More evidence of site-specificity in cinema's diegetic universe creates scope for mimetic representation. In comparison, dubbing has encouraged a lack of actual spatial information in filmic space. The inability of dubbing-based sound design to provide evidence of pro-filmic space has created somewhat 'unsitely' cinematic sound experiences. I have shown in this book why actors needed to perform twice: once on-location in front of the camera, and again in the studio in front of the studio microphone, where the actual situation of the site would be impossible to re-create.

As mentioned, this 'unsitely' or out-of-place mode of sound experience in dubbing-based film productions asks the listener to engage with the interior diegetic world of the cinematic work through memory and imagination – since there is not much site-specific sonic information provided about the various places depicted in the narrative. While encountering the film, the listener tends to construe a spatial reality that is often highly subjective. This subjective and improvizational quality of engagement is evident in the sound experience of a non-fiction film like *Aapothkalin Trikalika* ('The Kali of Emergency', Avikunthak, 2016). Ashish Avikunthak is a leading figure of Cinema of Prayoga and has been making films for the past twenty-five years. In *Aapothkalin Trikalika* many passages of sound don't have a mimetic connection to the site 'shown' on screen. The anticipated spatial characteristics of various sites such as a riverbed, a passageway under a bridge, an open field beside a factory, and inside a visibly empty room are largely absent from the sound. As the setting visibly changes from one scene to another, the auditory setting or the mise-en-sonore (Chattopadhyay, 2021a) doesn't change in the same way, as expected: an empty room sounds less reverberant than it should, a voice appears with added reverb in an empty field. These sonic incongruities, which sound 'faulty', are similar in many dubbed films, often hampering the work's credibility in relation to a realistic representation of site. But *Aapothkalin Trikalika,* like many other Avikunthak's works, unapologetically presents the post-synchronization practice of dubbing and recreating entire film soundscapes inside the studio as an apt methodology. Other contemporary Cinema of Prayoga filmmakers such as Amit Dutta, director of *Nainsukh* ('Joy of the Eyes', 2010) and *Sonchidi* ('The Golden Bird', 2011), among many other masterworks, Kamal Swaroop, author of cult films such as *Om-Dar-Ba-Dar* ('I'm Door-by-Door', 1988), and earlier pioneers such as Pramod Pati, use an asynchronous sound methodology. A variously devised dynamic relationship between audio and visual material has become a creative strategy for making subjective, poetic, and contemplative procedural filmmaking interventions. In the work of video or moving image artists, such as Kabir Mohanty, author of video installation works *Dwelling* (2006), *Handheld* (2009), and *In Memory* (2009, 2012), or Amar Kanwar, author of film installations *Scene of Crime* (2011) and *Such a Morning* (2017), such malleability of audio and video are methodological tools.

As Avikunthak suggests, he intends to disrupt and dissolve the modernist 'perspective' of cinema to foreground 'temporality rather than narrative in contemporary cinematic practice'. This dissolution helps pursue an ability to free the audience from an assigned (sonic and spatial) perspective, triggering a stream of thoughts and questions in the mind of the spectator/listener in the process. For him, then, post-synchronization practice is most suitable for conceptually and practically producing sounds that are unsitely, improvizational and transcendental – closer to the Indian epistemology of sound exemplified by Indian classical musical forms such as *Dhrupad* performances.

In contrast, synchronized sound practice demands a mimetic representation of site to consolidate the perspective brought in by the typically Western modernist approach, which advocates recording's progressive technological innovation. In sync sound-based films, the modernist idea of recording as a progressive method is glorified. For example, the camera, as a recording device, was the child of modernism, intended to record spatial perspectives as site-specifically and mimetically as possible, whereas recording was an alien practice to India's traditional oral cultures.

It is my assumption that this 'orality' is being revived in Avikunthak's work, catering to an Indian epistemology of sound. 'Orality' is considered here as a cinematic device where the voice or speech and spoken words are the primary narrative's hints and fragments. Rather than mimesis, the concept of quasi-diegesis is important here, as temporality unfolds through spoken narration, oral transmission, and suggestions (quasi- diegetic – arguably an aesthesis-driven Indian epistemology) rather than 'showing' (mimetic – essentially a logic and evidence-based European epistemology) what's happening on screen. *Aapothkalin Trikalika* underscores the pleasure embedded in this Indian epistemology of sounding and listening somewhat distilled in the *Dhvani theory* proposed by Ānandavardhana (c. 820–890 CE), the author of *Dhvanyāloka*. More focused discussion and elaboration along these lines of thought can be found in my upcoming monograph: *Sonic Perspectives from the Global South: Connecting Resonances* (2024) and other writings from my ongoing postdoctoral research (Chattopadhyay, 2023a, 2023b, 2023c).

The new generation of non-fiction filmmakers enriches the discursive sensitivity in their film-works. An artistic, context-aware, and nuanced approach to sound, a revealatory audio-visual relationship, and a dynamic audio-audio relationship appear to be central to their practice. Among them, Payal Kapadia, who won the Golden Eye award for best documentary film at the 2021 Cannes Film Festival for her non-fiction work *A Night of Knowing*

Nothing (2021) and competed at the same festival in 2017 with her previous work *Afternoon Clouds* (2017), states: 'For me, sound is the most important thing in cinema...The way I look at it, if you watch cinema with your ears closed, you won't feel anything. Why is cinema different from a painting? It's because of the passage of time. And time is measured through sound' (Dore, 2017). Shaunak Sen's award-winning film *All That Breathes* (2022) carefully, attentively, and affectionalely listens to the climate-affected black kites who fall from the sky disoriented with the smoke, fog and sounds of the evergrowing megalopolis Delhi reeling under environmental pollution. The unflinching sound of the city cannot drown out the desperate cries and deafening silences of the birds. The film resonates with theses dynamics and a great sensitivity.

References

Ānandavardhana (1990). *The Dhvanyāloka of Ānandavardhana with the Locana of Abhinavagupta*. Translated by Daniel Henry Holmes Ingalls. Cambridge, MA: Harvard University Press.

Battaglia, G. (2017). *Documentary Film in India: An Anthropological History*. London and New York, NY: Routledge.

Chattopadhyay, B. (2014c). "Sonic Drifting: Sound, City and Psychogeography". *SoundEffects* 3(3), 138–152.

Chattopadhyay, B. (2021a). *The Auditory Setting: Environmental Sounds in Film and Media Arts*. Edinburgh: Edinburgh University Press.

Chattopadhyay, B. (2021b). *Between the Headphones: Listening to the Practitioner*. Newcastle upon Tyne: Cambridge Scholars Publishing.

Chattopadhyay, B. (2023a). "An Equal Sound". *Rewire Festival Context Programme* 2023. www.rewirefestival.nl/feature/an-equal-sound (accessed 5 April 2023)

Chattopadhyay, B. (2023b). "Re-Sounding Souths". *CTM festival journal* – www.ctm-festival.de/magazine/re-sounding-souths (accessed 5 April 2023)

Chattopadhyay, B. (2023c). "Dhvāni: Resonance", in Herzogenrath, B. (ed.), *Concepts: A Travelogue*, pp 113–120. New York: Bloomsbury.

Chattopadhyay, B. (2024). *Sonic Perspectives from the Global South: Connecting Resonances* (working title, forthcoming).

Chion, M. (1994). *Audio-vision: Sound on Screen*. Translated and edited by C. Gorbman. New York: Columbia University Press.

Garga, B.D. (2007). *From Raj to Swaraj: The Non-Fiction Film in India*. Delhi; Penguin India.

Kaul, M. (1991). “Seen from Nowhere”. In Vatsavan, Kapita (ed.), *Concepts of Space: Ancient and Modern*. New Delhi: Abhinav Publications.
Rajadhyaksha, A. (2016). *Indian Cinema: A Very Short Introduction*. Oxford: Oxford University Press.
Ray, S. (1976). *Our films, Their Films*. Delhi: Orient Blackswan.

13. A Concluding Voiceover

Abstract: Critical listening, descriptive analysis, and informed commentaries combine in concluding remarks on the uniqueness of sound practices in India, their historical trajectories and aesthetic shifts. This final chapter locates the dominant tendencies and predilections of this diversity with a practice-aware approach. Critical and self-reflective observations from a sound practitioner's perspective drive this creative approach, identifying several complex and intercepting threads of production trajectories in a vastly heterogeneous national cinema, loosely unified by technological and aesthetic developments. A taxonomical model based on critical observations of historical and technological shifts and emergent aesthetic strategies provides the framework.

Keywords: Indian cinema, sound studies, media arts, sound production, film and media studies

In a Nutshell

Sound has received extensive academic interest in recent years. However, much of this attention has been invested in studying sound within an American and/or European media cultural context. Sound in other global regions, for example in India, has remained underexplored. And yet India is one of the world's largest producers of films and audiovisual media, with a formidable international presence as an emerging arts and cultural production giant. This book develops a comprehensive understanding and overview of the unique sound world of Indian film and audiovisual media to fill this void in analysis. It examines the historical developments of sound from early optical recordings to contemporary digital audio production alongside an aesthetic inquiry into mainstream, Parallel Cinema, and audiovisual art works. As a sound and media artist who studied film, I bring a practice-based methodology to these themes, incorporating extensive conversations with prominent sound practitioners and filmmakers active

Chattopadhyay, B., *Sound in Indian Film and Audiovisual Media: History, Practices, and Production.* Taylor & Francis Group, 2024
DOI 10.5117/9789463724739_CH13

in India. The book locates an emerging spatial and social awareness of film and media through sound, occurring alongside or against a traditionally transcendental approach to listening. Tracing this confluence of tradition and modernity, the book makes fresh contributions to the fields of film, audiovisual media, and sound studies.

The validity of the term 'Indian', housing a diversity of practices from Bombay and regional cinemas, is discussed. Despite linguistic and other vernacular differences, India's geographic unity, its common use of technology and technology transfer, and ongoing cultural exchanges within a methodological mélange suggest the term 'Indian cinema' is a useful, broad umbrella term.

Indian films are well known for producing transcendental sonic experiences, exemplified by the overwhelming use of non-site-based song-and-dance sequences. In this practice, the careful application and attentive organization of sounds is generally lacking from the narrative strategy. Indeed, there are many examples of popular Indian films and media productions where mindful sound design is not apparent; a loud, high-pitched auditory setting is often used instead to provide a remote and imaginary sonic landscape. This uncontested perception of Indian cinema needs to be challenged, however. If we consider the historical trajectories of Indian film sound production, we see that exporting an essentialist typecast is unnecessary and erroneous. Early Indian films used directly recorded locative sounds. And since the advent of digital technology in the 2000s, it has been possible to incorporate multi-layers of sounds in the production scheme and methodologies of Indian films and media works. A thorough study of this historical trajectory, via close readings of certain films and filmmakers, focusing on their creative use of sound, has helped place this inquiry, its analysis, and observations in context. A practice-based perspective in relation to the methods and approaches taken by sound practitioners provides the empirical evidence to qualify arguments. Extensive interviews and long conversations conducted with practitioners plus critical self-reflection add real-world knowledge.

I disclose how early Indian cinema's primarily vococentric approach to sound left little room for the creative use of other sound components. Nevertheless, filmmakers like Satyajit Ray, Ritwik Ghatak, and Mrinal Sen attended to the detail of sites, people, habitats, environments, and situations, narrating India's socio-political realities through situated listening. This specific, realistic sensibility tended to overcome the limitations and expand beyond the creative possibilities of direct or synchronized sound recording and monaural mixing as practiced in Indian cinema at that time.

Technically informed dubbing and studio-centric analogue processing dominated Indian popular film sound production between 1960 and 1990. These processes rendered action and song-and-dance sequences in foreign locations as something spectacular. The resultant cinematic experience, based on emotive tension and dramatic stimulation with few nuances, provided a sense of escape from the troubled sites of India's social realities.

Parallel and art house practitioners, women filmmakers, and Cinema of Prayoga artists have explored the same sound technologies and methods used by their mainstream contemporaries, but they have pushed the normative margins of the craft. Investigating their creative use of sounds that convincingly portray pro-filmic spaces in narration, I suggest that these parallel filmmakers developed, to varying degrees, a tradition of creative sonic sensibility, which independent filmmakers have tried to resurrect in the digital era.

Historiography of Indian Sound Production Practices

Critically listening to sound production trajectories within Indian cinema exposes how site-specific auditory realities and sonic spaces have been inconsistently rendered and produced through various phases of sound production. I have divided these trajectories into three primary historical markers that seem useful as a heuristic method for locating and remapping the foremost technological and aesthetic shifts. I have studied how these shifts manifested in the emerging aesthetic choices available to and embraced by sound practitioners, and how these choices are reflected in the sound production of Indian films. In other words, the various forms and formats of technological innovations and transformation have influenced the degree of spatial presence produced through the creative use of sound components. Therefore, my central concern has been to study the specific nature of creative sound practice in these three corresponding and intercepting phases of sound recording and production in Indian cinema: 1) optical films and direct sound recording, analogue magnetic tape recording, and monaural mixing; 2) dubbing or ADR, and studio-centric stereophonic mixing; and 3) digital multi-track synchronized/sync recording and surround sound design. Critical observation, reflection, and analysis of Indian films from these phases of sound production qualify an evidential account of the research following a bottom-up approach. The book questions whether this approach of taking a broad historical overview suffers from not being able to thoroughly delve into a specific era of Indian cinema. In its defence, I refer to the argument of film and media historians who maintain that writing history demands a broader perspective to accommodate historiographical accounts – an argument I have discussed and elaborated in my earlier book *The Auditory Setting* (2021a) based on my studies of environmental sounds in global cinemas. Emerging from a lack of serious and sustained research on sound in Indian films, this book, with its inherent constraints, creates a context for future research that might present more detailed studies of specific periods of sound production in India. Moreover, this book has been inspired by my own professional background as a sound practitioner and academic experience related to the historical developments of sound in Indian cinema, opening the research up to a new range of auto-ethnographic interventions and self-reflexive knowledge generation. As film historian Casper Tybjerg argues: 'In order to connect the dots of the historical record into some sort of coherent pattern, historians must inevitably draw on their own experience and understanding of human life and behavior...[we] must inevitably rely on our background knowledge of the world when we discuss the past' (2013, n.p.).

The following tripartite model first appeared in the concluding chapter of my other book on Indian cinema interwined with this book's research: *Between the Headphones* (2021b), a complementary collection of interviews and conversation with leading Indian sound practitioners, and a reference volume of fieldwork useful for this book. Here these outlines are underlined again for the purpose of consolidating my analysis and critical historicizing of the evolution, development, and trajectories of sound practice in India.

1. Early Developments and Direct Sound Aesthetics

From 1931, when sound burst onto the Indian scene with Ardeshir Irani's film *Alam Ara*, until the late 1950s, Indian cinema, largely comprising of music-oriented and/or devotional content, adapted to the technicalities of direct synchronized sound. Throughout this trajectory of direct sound use, characterized by recording techniques and equipment with a somewhat limited dynamic recording and production range, the use of microphones on location was mostly directional and focused on what Chion also diagnoses in American cinema as 'almost always the voice' (1994, p. 5), establishing sound films as an essentially vococentric order. The directly recorded sound in these films provided a predominantly monophonic narration of the pro-filmic space. Occasionally, louder 'unwanted' sounds intruded as audio leakage onto the film's soundtrack, providing spatial information about the pro-filmic space being narrated in the film.

The significance of such sound elements lies in them being derived from the real location where the film was shot. These site-specific sounds provide distinctive and realistic evidence of a live pro-filmic space registered in early direct sound practices in Indian cinema. Social realist filmmakers, who purposefully represented locations, settings, spaces, and situations, recognized and emphasized sound's role in locational observation and documentation, establishing a legacy for realism in Indian cinema. Here 'realism' refers to the tradition of observational cinema that represents reality by recording vision and sound that 'comes from within the world of the film' (Kania, 2009, p. 244).

Satyajit Ray's debut film, *Pather Panchali* (1955), for example, enables its audience to relate to different locations of the village where the film was shot. This was achieved via the incorporation of actual site-specific sounds such as wind through grasslands, the drone of electrical poles beside the railroad, the friction of tree branches blowing in a gentle breeze at the forest's edge, etc. In *Aparajito* (1956), also part of Ray's *The Apu Trilogy* (1955–59), one can distinctly hear different urban areas of Benares through the ears of protagonist Apu following his exploration of the city. The respective cinematic passages are built with a spatial use of sounds that captures their location-specific textures, realistic depth-of-field, and detailed perspectives. In *Charulata* (1964) Ray's elaborate use of sounds from street sellers and their antics engage the film's audience with its representation of a secluded and idyllic 1870's Calcutta neighbourhood. Such a spatio-temporal manifestation of sound was one of Ray's challenges to populist Indian cinema of his time, which was otherwise a typically verbose and vococentric exercise reliant on character dialogue as a primary source of narration, arranged within a loud musical score, sporadically punctuated by loosely arranged songs.

2. The Dubbing Era and Studio Centric Sound Production

Direct recording continued throughout the 1950s. The period, known as the Golden Age of Indian cinema (from the late 1940s to the late 1960s), when middle-ground filmmakers such as Bimal Roy, Chetan Anand, and Guru Dutt emerged and placed their national cinema on the world stage, bridged the transition to more convenient, portable, and robust magnetic recording and re-recording. Studio-centric production practices developed from the 1960s onwards, following the commercialization of popular Indian films. The subsequent era is renowned for its colourful antics, half-recognizable foreign locations, and spectacular song-and-dance sequences, largely produced in the 1980s and 1990s, yet still somewhat ongoing today.

When the magnetic medium emerged, it became possible to clean, erase, overdub, and employ multi-track mixing. Indian cinema's use of loop dubbing and ADR became a regular practice when Arriflex 2C and Arriflex 3 cameras became available from the late 1960s, which required a blimp (a soundproof cover) to reduce their notorious motor noise during location shooting. Such distracting camera noise meant that all sound had to be re-created in the studio. Eventually, this practice became standard in Indian films, dominating Indian cinema roughly from the 1960s to the late 1990s.

This long stretch of time illustrates a persistent interest in the controlled deployment of a few sound elements as design material in films, keeping the primacy of the voice alongside prominent background music, song-and-dance sequences, and processed sound effects. In this hierarchy of sound organization, locative sounds were notably lacking. Sound practitioners who used dubbing practices had to accept that the narrative strategy of post-synchronization focused on a clean and legible vocal sound over other components. Studio technicians constructed an artificial film soundtrack, using songs and loud background music to mask weaknesses in sound recording and mixing. This technologically mediated approach was predisposed to overly expressionistic, spectacular, and melodramatic representations far removed from the reality and authenticity of a film's pro-filmic space, rendering an expanded fantasy-like experience that proved popular with mass audiences.

3. The Digital Realm and Emerging Sonic Realism

During the late 1990s to early 2000s, easy-to-handle digital hard disc-based recording devices with larger storage, audio workstations, and sound design software became widely available in India, heralding a gradual yet large-scale conversion away from analogue sound recording and post-production practices, and optical film exhibition. Sync sound recording techniques and surround sound formats enabled significantly larger dynamic ranges, larger headroom, multi-channel formats, wider panning for sound spatialization, full-frequency channels (20 Hz–20 KHz) with a flatter response, and increased complexity in mixing and spatial fidelity, which have opened multiple creative possibilities to film sound practitioners (Holman, 2002; Kerins, 2011). Innovative uses of digital sound in Indian cinema have led to new experiences in which audiences can engage with more realistic representations of situations, locations, and people, often through situated listening, involving the novel spatial organization of sound.

The ramifications of Indian cinema's adaptation to digital technology have been far-reaching. New perceptions of how sound can be used have been especially evident. Digitalization's substantial impact on cinema's production/reproduction chains has had a consequent effect on the aesthetic choices, strategies, and resultant appreciation of sound as a creative medium for artistic intervention and innovation. Digital sound technologies have altered the way in which film soundtrack standards are perceived; even routine song-and-dance sequences from mainstream Indian films have had to be reconfigured in the digital era.

Likewise, digital technologies have reordered the organization of sound and the sonic environment in Indian films. Sync sound now allows for sound to be recorded on location in synchronization with the live performances of the actors in front of the camera. Digital recording devices enable access to the farthest corners of locations with precise control over every clip. They also have ample storage for multiple tracks. Site-authentic sound recordings – mostly of actors' voices recorded live in situ, which include aspects of their direct setting such as actual sound reflected from walls and other surfaces, site-specific ambience, and sound effects – remove the need to add extensive stock sound effects and pre-recorded ambience from external sound banks in post-production. The use of situated sound recordings in the spatial design process has opened up in-depth methods and options for translating pro-filmic spaces and their embedded social, anthropological, geological, environmental, and non-human narratives. In the studio, many choices exist for digitally enhancing location-specific actual sounds, edited and cleaned with noise filters such as Dolby NR, in preparation for multi-track surround design. Various applications are now available that can manipulate recorded sounds to restructure and reorder their spatial characteristics (e.g., reverb) for a more precisely designed cinematic sound experience. Indeed, sound production aesthetics have significantly been redefined and advanced terms like 'sound design' have appeared.

As the new trend in sync sound and surround sound design becomes the standard experience anticipated in emerging urban Indian multiplexes and OTT platforms, sound practice incorporates the newly available technological improvements over existing set-ups. Fast technological developments impact both sound post-production techniques such as non-linear editing, collaborative designing, and re-mixing in multi-channel studios, and how sound is replayed in new film theatres and multiplexes, or via streaming networks. Sound is either mixed in Dolby 5.1 and 7.1, the recently introduced Auro 3D, or Dolby Atmos. All these formats work with technologies that split sound into multiple surround speakers. Production practices involving digitally recorded and organized sounds are employed in novel modes of cinematic expression. They have evolved and reconfigured the earlier linear construct of a soundtrack into a spatially evocative surround sound environment. Layers of location-recorded sounds lead viewers/listeners to renegotiate their association with the filmic experience through perceptual processes involving spatial cognition, mimesis, presence, and immersion. Besides discussing these new developments, this book also draws attention to the inherent limitations of an industrial sound practice lacking artistic sensibilities in popular mainstream films aimed at making easy profits with little investment. As a contrasting parallel, I point to how portable, easy-to-use digital recording devices have accelerated audiovisual media arts. New forms of sound practices have emerged due to the availability of technologies and access, via the internet, to exploratory works such as field recording-based sound works, and electronic and experimental music, etc.

However, artistic sensibilities are also being explored in certain aspects of Indian cinema in the digital era. Sound as a storytelling and cinematic communication tool has become central to Indian cinema, especially in independent films, experimental, and non-fiction films. Whereas sound often took a back seat in relation to the pictorial quality of images in the erstwhile analogue era, it now has a more prominent role in engaging and immersing audiences. Site-specific, incidental, ambient, and all actual sounds are considered indispensable to today's Indian cinema.

Drawing Out Sonic Aesthesis

Indian cinema – with its diverse fields of practices, productions, and experiences – is one of the world's largest producers of films and audiovisual content. This book locates the dominant tendencies and predilections from this diversity via a practice-led approach. Methodologies of critical and self-reflective observation reveal several complex and intercepting production trajectories in a vastly heterogeneous national cinema, which is nonetheless unified geographically, technologically, and aesthetically. This observational approach has drawn out Indian cinema's major historical developments and transitions with a particular focus on the creative use of sound and the evolving sonic aesthesis: namely, how audio-audio relationships are handled in different production situations and contexts; how nuanced approaches are conceived and realized within technological constraints and industrial structures through processes of diegesis and mimesis; and how sensitive sound components such as site-specific sounds are dealt with to enhance the sense of presence. These are underexplored subjects in film sound research, and within the wider field of Sound Studies. Through these inquiries, this book reveals the aesthetic strategies and approaches to sound practice and production in Indian cinema.

This aesthetic inquiry has a particular interest in spatial practice, one in which sound helps to expand the spatiality of the cinematic experience. As discussed in the Introduction, 'spatial turn' is understood as 'the increasing recognition of the intimate links between sound and space' (Eisenberg, 2015, p. 195), denoting a focused discussion on the notion of place and space in the humanities. Responding to the idea of the spatial turn, I have traced the sonorous space's evolution in Indian cinema. The transformation of the diegetic space from a screen-centric, monaural soundtrack via the ultra-screen expanded stereophonic space to today's spatially enveloping surround sound environment can be understood as paradigmatic shifts from 'looking at' to 'being in' (Dyson 2009, p. 2) – a clear indication of a spatial shift characterized by the changing relationship between site, space, and sound as created through evolving production practices. Much of this shift is made audible through the use and spatial ordering of sounds to create an immersive environment, facilitating an embodied experience of spatial fidelity and presence. I have shown how the screen-centric monaural recording and mixing of synchronized and locative sound as direct evidence has helped to 'trace the site'. Dubbing, ultra-screen stereophonic mixing, and their deliberate lack of site-specific sounds create an auditory setting of an unsitely spectacle to 'escape the site'. I have also demonstrated that the contemporary digital era is more generous in including more spatial and situated sound elements in the sound organization than former eras. Consequently, the site becomes more bodily 'present'. I have analysed significant passages from Indian films after 2001, showing that many embraced digital, multi-track sync sound recording before it became standard practice through the first decade of the 2000s. The novel experience of listening to the filmic space in the digital era is marked by low frequency room tones and other atmospheric content recorded in synchronization with the on-site location. The elaborate spatialization of these sound recordings render an auditory setting, which provides ample evidence of the site in a spatially enveloping surround sound environment. These new methods and approaches produce a sense of 'being sited' in Indian sociopolitical realities as an embodied experience of pro-filmic space, of its convincingly palpable presence. The transitions within sound production, from analogue eras to the contemporary digitalization of sound, show how the latter period has triggered novel spatial experiences of more-than-immersion, embodiment, and site-oriented engagement in Indian cinema. It is widely acknowledged that film is possibly 'the first medium to efficiently accommodate sound' (D'Escriván, 2009, p. 65), and my aim has been to uncover how sound practices in Indian films reveal locative realities. In other words, in this book, I have studied the historical trajectories of creative sound practice in India, and how emergent sensibilities in different phases of this practice have manifested in subcontinental films and audiovisual media to enhance spatial experiences and place-making. In-depth conversations with sound practitioners have informed and motivated a practice-led direction in the book. I have been critically listening to how they suggest that cinematic spatiality is crucial to ground the audiovisual experience in reality, and produce a sense of credibility, orientation, and believability.

Reassessing Ray, Audiographic Realism, and Independent Creative Practices

Satyajit Ray began writing film reviews very early on in his career. In 1948 he published a short but perceptive commentary entitled 'What is Wrong with Indian Films'. The essay criticized the predominance of saccharine sweet musicals and religious mysticism in the Indian cinema of his time:

> The raw material of the cinema is life itself. It is incredible that a country which has inspired so much painting and music and poetry should fail to move the moviemaker. He has only to keep his eyes open, and his ears. Let him do so. (1976, p. 24)

The essay expresses his polemical position and critical reading of cinema that existed far from lived environments and their many realities. Ray asked filmmakers to keep their eyes and ears open. J.P. Maxwell's observations on film could be applied to Ray's films too, which 'simulate the perceptions of an observer located on the film set, whose eyes and ears (camera and microphone) are joined as inseparably as those of a real head' (Lastra, 2000, p. 183) – an essential tenet of perceptual realism provided by synchronized sound and monaural mixing practices.

However, as shown in this book, this technological framework had already been practiced in Indian cinema since the advent of sound in *Alam Ara* (Irani, 1931), continued when analogue quadrophonic/stereo was invented and applied in path-breaking films like *Sholay* (Sippy, 1975), and maintained until the early 1990s when many films turned to stereophonic mixing. Why then is Ray significant, referential, and exemplary in this context? When studying the trajectory of synchronized sound and monaural aesthetics, Ray provides a key reference point because his work exploits the territories of sync sound and pushes the boundaries of aesthetic choices that address contemporaneous social-political realities. The sound in his films is a benchmark in practicing 'audiographic realism'. It considers the site and respects the pro-filmic space, triggering a locative, embodied sonic experience derived from reality, which is not in any way escapist or fantasy-like. In 1951 Ray wrote:

> For a popular medium, the best kind of inspiration should derive from life and have its roots in it. No amount of technical polish can make up for artificiality of theme and dishonesty of treatment. The Indian filmmaker must turn to life, to reality. (1976, p. 127)

This statement aptly frames Ray's realist aesthetics, which most of his contemporaries working in the Indian mainstream evaded. The same standard monophonic system in their hands remained primarily vococentric, dominated by the rhetoric of narrative pleasure with the normative structure of song-and-dance sequences, celebrating non-diegetic fantasy spaces.

Ray expanded his vision of the monophonic system to create a window on the world formed within the home. Audience members sonically develop an 'enormous curiosity about the world' (Ganguly, 2001, p. 63) through such a window. This curiosity leads to experiencing an opening to an outside within the mental synchronization of sound in monaural films (Chion, 1994, p. 69). Indian film scholar Supriya Chaudhuri argues that Ray did 'provide us with a cinematic staging of interiority that might compensate for the hollowness of the house. Such interiority can only be located in the exterior, in the open, empty world, in a mental rather than a physical place' (2007, p. 12). In this process, Ray allows his audience to respect the worldly sites from where the stories of real people unfold and takes these sites as the point of departure for producing a richer extended cinematic experience based on reality. This specific sensibility makes Ray's authorial positioning clearer. Shyam Benegal describes Ray's position within Indian cinema as a 'splendid isolation', adding that 'I locate Indian cinema as before Ray and after Ray' (Chattopadhyay, 2021b, p. 30).

Ray's directorial debut, *Pather Panchali* (1955), broke away from conventional cinema of the 1940s and 1950s. It guided cinema-going audiences into the real world, where poverty, empathy, tragedy, and comedy sensitively combine without resorting to sentimentalism, which was then prevalent in Indian cinema. Ray's worldly sensibility replaced the popular domestic melodrama's theatricality between characters, whose dialogue ignored the sitely evidence of sound. Ray also challenged the perpetual intervention of the non-diegetic musical score with song-and-dance sequences, which had been the dominant mode of narration. Ray's creative use of sound, especially ambience and other actual sounds, established a diegetic presence of the real world, recognizing and giving prominence to the pro-filmic space in the narration, revealing beauty in the everyday. His works delineate the possibilities of narrating and revealing the intricate details of lived environments and their inhabitants – the people and their multitude of realities – for future generations of Indian filmmakers and sound practitioners.

Though emerging from the synchronized sound tradition, which was practiced at the very beginning of Indian film sound in the 1930s, Ray's sonic sensibility received inadequate recognition during his lifetime notwithstanding his influences on the younger generation of parallel filmmakers such as Shyam Benegal and Adoor Gopalakrishnan. His approach is finding a form of revival in Indian cinema's digital location sync sound recording era. Digital sound practices facilitate the creation of realistic auditory settings that relate site-aware experiences informed by socio-spatial awareness.

An Emergent Sonic Spatiality in Indian Cinema

A significant proportion of this book's investigation into sound production processes centres on synchronized sound recording. Terms like 'location sync sound' or 'live sound' are used in the film practitioner's vocabulary to refer to the recording of primarily the voice on location in synchronization with the live performance in front of the camera instead of re-recording it inside the studio in post-production dubbing sessions. The digitalization of post-1990s film technology represents a cinematic revival of observational techniques related to location-specific detail, which also includes ambient sound and sync sound effects. An emergent fascination with authentic location and spatial evidence of the film's image suggests the rediscovery of cinema's realist roots. These 'actual sounds', which Ray describes as a means to 'create change in mood' rather than music (1989),[1] correspond to

1 Satyajit Ray in an interview with Pierre Andre Boutang states, 'I can use actual sounds creatively to serve the purpose of music and to allow a certain change in mood to be perceptible to the audience', which I used as a key statement in this book.

the site and the situated sensation of 'being there' (Chattopadhyay 2016, 2021a) by enhancing a sense of sonic spatial presence.

The book reveals how sync sound brings with it a certain spatial credibility that helps to convince the audience of the story's diegesis. Direct sound produced during the analogue era worked with similar principles of recording a synchronized voice but was rendered in a screen-centric monaural setting, thereby limiting the scope of the spatial diffusion and contiguous cognition of site-specific realities. Moreover, the limited dynamic ranges of analogue recording (optical and magnetic) couldn't register essential low-frequency room tone or other spatial auditory presences. Digital technologies overcome these limitations. Multi-layered and richly evocative audio recordings, collected from the place of narration and built into a digital surround sound cinema environment, plays out a spatial topography of the pro-filmic space in the minds of its audience. This, in turn, allows the audience to immediately recognize and relate to the constructed cinematic space. At the reception end, this specific practice of sound manifests itself in novel modes of audience cognitive association that differ substantially from monaural or stereophonic cinematic experiences, reconfiguring the linear construct of a 'soundtrack' towards a spatially evocative surround sound environment. Digital surround sound's authenticity in the cinema is also distinct from other emerging media sound practices in contemporary India such as video games and new media artworks, where questions of space and place are addressed by subjective, imaginary, and inventive means. Today's synch recording and surround sound techniques, in their various independent and creative digital cinema uses, compel audiences to engage with the sensorial and the phenomenal through a spatial cognition of sound.

While my principal findings in this book may resonate with the digital era's promise of an unmediated merging with reality, my research also demands a closer, critical look at the concept of presence. Spatial presence in Indian cinema emerges from a functional approach to mimetically (re) presenting sound's inherent site-specificity. In *The Auditory Setting* (2021a), I suggest that mediated presence is often manufactured, technologically crafted, and constructed, rather than being an immediate, sensitive, and direct exploration of a setting, particularly in commercially focused, popular films aimed at mass entertainment. The many layers of the real India and its nebulous but wonderful places and sonorous spaces, not to mention the situated social, cultural, political, religious, environmental, and other profound issues Indian rural hinterlands and urban spaces continue to have since the arrival of sound in cinema, are rarely sonically 'present' in film. The truthfulness of these realities withdraws or slips away under the pressure of providing storytelling pleasure and cinematic entertainment, conditioned by the normative and industry-prescribed use of technology, avoiding taking any aesthetic risk. In my opinion this habitually elusive spatial veracity in Indian films is primarily determined by the pervasive control of the market and the film industry at the expense of greater artistic freedom. The sound practitioner's sincere creative efforts have historically met with resistance and hostility arising from an adherence to, or the demand to adhere to, the norms of the film industry.[2] This has led to inaccurate and untruthful (re)presentations of reality in favour of a manufactured presence, what film scholar Stephen Prince has called 'perceptual realism', meaning that unreal images (and sounds) appear 'referentially fictional but perceptually realistic' (1996, p. 32). The apparent intensification of presence in Indian cinema in the digital era is largely artificial and constructed – with the help of market-driven digital technology with its readily available tools, presets, formats, and systems – rather than tapping (or hacking) into the contemporary, post-digital[3] open-source playing field with its immensely creative applications available to filmmakers, producers, production houses, studios, and guilds.

On many occasions, the dense and noisy parts of site-specific sound recordings are controlled and sanitized through editing and advanced noise reduction to provide cleaner sonic textures, whereby more sterilized accounts of sonic spaces and environments are heard. Altering sound with pre-set effects such as reverb and EQ in sound production is a conscious process of manipulating sounds to control and compensate for a lack of realistic sonic representations. This film industry-driven technological approach to sound production can be understood within Indian cinema's asynchronous tendencies: separating sound from its real, site-specific sources enhances its clarity and legibility, as well as dramatic and spectacular qualities. This perspective on deliberately processing sounds seems comparable to what sound scholar Emily Thompson has articulated within the context

2 In *Between the Headphones* (Chattopadhyay, 2021b), Indian sound practitioners have expressed their frustration with not being able to contribute their creative best. To this question of artistry and nuance, and the suggestion of instigating change, or voicing their opinion, they have responded with sarcasm, irritation, withdrawal, or hostility. While their defence of their profession and their precarious position as artists and craftspeople in the system is understandable, it is also imperative to pay attention to what they truly aspire to, revealed through the intermittent flashes of inspiration with which they describe their work.

3 When digital sound production has reached saturation point, contemporary media theories suggest that a critical juncture will occur in which old and new media will merge to form hybrid methodologies and aesthetic conditions. Such a 'post-digital' context would direct researchers to explore potentially new fields of inquiry regarding sound practices.

of American cinema as 'a fundamental compulsion to control the behavior of sound' (2004, p. 2). This compulsion for achieving clarity in the cinematic soundscape has led the practitioner to often employ easy and partial 'sound-marks' instead of listening in-depth to and registering a place's situated and authentic atmosphere. This tendency to highlight a stereotypical sound, often at the expense of the many other actual sounds emanating from a public site in India, is meant to balance out the noise reduction and editing of digital sync sounds during post-production processes. These industrial norms, practical rules, and regulations embedded in the essentially functional and marketing aspects of film sound production tend to hinder artistic potential of the sound practitioner and often fail to further enrich the film's spatial features, especially in mainstream, industrial films from Mumbai and the regions. As a result, the multilayered 'public-ness' of Indian locations remains mostly absent in mainstream film. However, sound takes a more central role within spatial cognition, and as a nuanced tool of artistic practice in Indian independent and art house films.

This discussion helps to enrich a specific area of sound scholarship, notably the spatial dimension of sound's relationship to public sites and the idea of spatial presence.

Towards a Qualitative Shift in (Film) Sound Production

This book outlines the historical trajectories of sound practice in Indian cinema based largely on the insights of its practitioners. Such a practice-led scholarly approach reflects the critical observations of technological shifts and emergent aesthetic strategies. Its point of departure is to be found in specific phases of technological innovation and transitions in sound production but is not limited to a discussion of sound technology's history. On the contrary, this situated inquiry highlights characteristics that delineate the sonic aesthesis emerging from these prominent technological phases. Delving into the essential elements of sound production through in-depth conversations with practitioners provides first-hand information of what has historically transpired in Indian cinema. And this research becomes a valuable prerequisite to understanding how sonic space has been produced and shifted in different phases of practice. A discussion of current practice and the claim that digital sound production has impacted the cinematic experience, making it more spatially elaborate, has been met with complementary empirical evidence and anecdotal accounts from conversations with practitioners, who also hint at the idea of a better practice, one in which their artistry is recognized as part of a qualitative shift.

There is no official document or manual related to best practice in sound production for Indian film industry yet. However, research done in this book, in my previous book (2021a), and insights found in my conversations with practitioners (2021b) may shed light on the perception of a better practice, critically gauging industry standards within this context. This is often reflected in national awards given to the 'best' works in categories for 'Location Sound Recordist', 'Sound Designer', and 'Re-recordist of the Final Mixed Track' on a national level each year in India.[4] These awards are based on the film industry's evaluation of the highest level of craftsmanship in sound production. But how do these 'best' works sound? Do they indeed represent and exemplify exceptional works of film sound production, those that demonstrate a sensitive application of artistry? As this book suggests, these awards do recognize the artistic and creative merits of certain works, and thereby inspire the community of sound practitioners. However, in both *The Auditory Setting* (2021a) and *Between the Headphones* (2021b), I have discussed in detail that sound-based creative endeavours are often characterized by a refusal to be standardized, destabilizing existing systems of industrial norms and protocols. To articulate how the idea of producing better sound occupies the minds of practitioners and how they aspire to achieve certain (personal) standards of quality and efficacy within the immense constraints of the film industry, I refer to the interviews that discuss how the creative utilization of site-specific sound has expanded in the digital era. It is actual location sound that is categorically singled out by these established practitioners as a primary element of artistic exploration in film sound production. However, sound production in mainstream Indian cinema is still dominated by the pervasive norms and rules of the film industry, even though the digital realm opens up possibilities for creative intervention by the practitioner, shaking up the hierarchical and feudal chains of industrial and studio-centric production. One example of this is how sync sound puts the actor's participation on the film set on a par with the location sound technicians, such as the boom operators,[5] who have long held a lower status in film crew hierarchy. In this book, I once again argue that the lenses provided by artistic freedom practiced in audiovisual

4 See 67th national film awards: www.thehindu.com/entertainment/movies/67th-national-film-awards-complete-list-updating/article34131921.ece (accessed 5 April 2023).

5 A boom operator works with the production sound mixer and is responsible for microphone placement, located on the end of a long boom pole.

media arts can be useful in the attempt to find out how film sound can be more sensitive towards spatial realities, more playful, and more aware, as well as more nuanced in its application. In conversation, Resul Pookutty hints at how an artistic sensibility could enrich the generally mechanized film sound production practices (Chattopadhyay, 2021b) – as it does in his case. Here I intend to distance myself from Indian cinema's industrial norms and set regulations in search of more artistic freedom, hacking the technology, and subverting industrial standards. I am, therefore, critical of the standardizing idea of a best practice when it comes to individual artistry and send out a call for greater inclusiveness and sensitivity to the spatial aspects of sound and a socially resonant creative exploration of sound in film and audiovisual media beyond normative models, industry regulations, and standardized formulas. Better practices, as I employ the concept here again, after I did so in *The Auditory Setting* (2021a), envision a future of film sound in India where these artistic and ingenious impulses will be explored, resonating with the situated social realities, expanding the industry-dependent ideas of the best sound in film only as a point of departure to reinterpret and recontextualize the conventional notion of best practice pre-sets towards emergent and continually improving exploratory sound practices.

I have suggested in this book that it is in independent and art house cinema, the Cinema of Prayoga, and audiovisual media arts, rather than in cinema for the masses, where the possibilities for artistic exploration and developing exceptional examples of creativity in sound are most present. Sound produced by a handful yet growing number of Indian independent filmmakers is taking its own course, creating layers of multiple impressions within, around, and beyond the visual narrative and the overarching story. Here the authors' (both the director's and the sound practitioner's) subjective interpretations of places, people, situations, and spaces are paramount and crucial when developing an evocative auditory setting.

Drawing also on the work of other non-mainstream filmmakers and women directors, better practice has the potential to foreground collaboration, camaraderie, fellow-feeling, conviviality, and community building rather than exploitation and profit based on marginalization and feudal hierarchies, prevalent in industry film productions. In this light, better practice has the capacity to reimagine a future of sound practice in India dedicated to artistic freedom, allowing practitioners time and space for sensitive listening, and to register their self-empowered positions within the field of sound.

References

Chaudhuri, S. (2007). "Space, Interiority and Affect in Charulata and Ghare Baire". *Journal of the Moving Image* 6. http://jmionline. org/article/space_interiority_and_affect_in_charulata_and_ghare_baire (accessed 3 July 2022).

Chattopadhyay, B. (2016). "Being There: Evocation of the Site in Contemporary Indian Cinema". *Journal of Sonic Studies* 12. https://www.researchcatalogue.net/view/286592/286593 (accessed 10 September 2023).

Chattopadhyay, B. (2021a). *The Auditory Setting: Environmental Sounds in Film and Media Arts*. Edinburgh: Edinburgh University Press.

Chattopadhyay, B. (2021b). *Between the Headphones: Listening to the Practitioner*. Newcastle upon Tyne: Cambridge Scholars Publishing.

Chion, M. (1994). *Audio-vision: Sound on Screen*. Translated and edited by C. Gorbman. New York: Columbia University Press.

D'Escriván, J. (2009). "Sound Art (?) on/in Film". *Organised Sound* 14(1), 65–73.

Dyson, F. (2009). *Sounding New Media: Immersion and Embodiment in the Arts and Culture*. California: University of California Press.

Eisenberg, A. J. (2015). "Space". In Novak, D. & Sakakeeny, M. (eds), *Keywords in Sound* (pp. 193–207). Durham: Duke University Press.

Ganguly, S. (2001). *Satyajit Ray: In Search of the Modern*. New Delhi: Indialog.

Holman, T. (2002). *Sound for Film and Television*. Boston: Focal Press.

Kania, A. (2009). "Realism". In P. Livingston and C. Plantinga (eds), *The Routledge Companion to Philosophy and Film* (pp. 237–247). London and New York: Routledge.

Kerins, M. (2011). *Beyond Dolby (Stereo): Cinema in the Digital Sound Age*. Bloomington: Indiana University Press.

Lastra, J. (2000). *Sound Technology and the American Cinema: Perception, Representation, Modernity*. New York: Columbia University Press.

Prince, S. (1996). "True Lies: Perceptual Realism, Digital Images, and Film Theory". *Film Quarterly* 49(3), 27–37.

Ray, S. (1976). *Our Films, Their Films*. Delhi: Orient Blackswan.

Thompson, E. (2004). *The Soundscape of Modernity: Architectural Acoustics and the Culture of Listening in America, 1900-1933*. London: The MIT Press.

Tybjerg, C. (2013). "The Sown and the Waste – Or, the Psychedelic Writing of Film History". *NECSUS_European Journal of Media Studies* 2(2), 509–525.

Bibliography

Allen, R. (2009). "Pather Panchali". In Lalitha Gopalan (ed.) *The Cinema of India*, 24 Frames Series (pp. 86–95). London: Wallflower Press.

Altman, R. (ed.) (1992). *Sound Theory/Sound Practice*. New York: Routledge.

Ānandavardhana (1990). *The Dhvanyāloka of Ānandavardhana with the Locana of Abhinavagupta*. Translated by Daniel Henry Holmes Ingalls. Cambridge, MA: Harvard University Press.

Arnold, A. (1991). *Hindi filmī gīt: On the History of Commercial Indian Popular Music* (Ph. D. Dissertation). Urbana-Champaign: University of Illinois.

ArZan (2019, October 14). "Ardeshir Irani, the Father of Indian Talkies who had many other Milestones to his Name". *Parsi Khaba*. https://parsikhabar.net/film/ardeshir-irani-the-father-of-indian-talkies-who-had-many-other-milestones-to-his-name/21171/ (accessed 5 April 2023).

Balázs, B. (1985). "Theory of the Film: Sound". In Weis, E. and Belton, J. (eds), *Film Sound: Theory and Practice* (pp. 116–125). New York: Columbia University Press.

Battaglia, G. (2017). *Documentary Film in India: An Anthropological History*. London and New York, NY: Routledge.

Biancorosso, G. (2009). "Sound". In Plantinga, C. R. and Livingston, P. (eds), *The Routledge Companion to Philosophy and Film* (pp. 260–267). London: Routledge.

Birtwistle, A. (2010). *Cinesonica: Sounding Film and Video*. Manchester: Manchester University Press.

Balázs, B. (1985). "Theory of the Film: Sound". In Weis, E. & Belton, J. (eds), *Film Sound: Theory and Practice* (pp. 116–125). New York: Columbia University Press.

Barthes, R. (1977). *Image, Music, Text*. Translated by Stephen Heath. London: Fontana.

Bhatt, O. (2019). "Writing about Sound: The Early Talkie Film Periodicals of India". In *International Perspectives on Publishing Platforms* (pp. 45 – 81). London: Routledge.

Bhatyacharya, T. (ed.) (2009). *Dhanimoyotar Itibritto*. Kolkata: Roopkala Kendro.

Bloom, P. J. (2014). "Sound Theory". In Buckland, W. and Branigan, E. (eds), *The Routledge Encyclopedia of Film Theory* (pp. 430–434). London: Routledge.

Booth, G. D. and Shope, B. (eds). (2013). *More Than Bollywood: Studies in Indian Popular Music*. New York: Oxford University Press.

Bordwell, D. and Thompson, K. (1997). *Film Art. An Introduction*. New York: McGraw-Hill Publishing Company.

Born, G. (ed.). (2013). *Music, Sound and Space: Transformations of Public and Private Experience*. Cambridge, U.K.: Cambridge University Press.

Borges, J. L. (1999). *The Total Library: Non-Fiction 1922–1986*. London: Penguin Publishers.

Blesser, B. and Salter, L.R. (2009). *Spaces Speak, Are You Listening?: Experiencing Aural Architecture*. London: The MIT press.

Bordwell, D. (2009). "Cognitive theory". In Livingston, P. & Plantinga, C. (eds), *The Routledge Companion to Philosophy and Film* (pp. 356–365). London: Routledge.

Bordwell, D. and Thompson, K. (1997). *Film Art: An Introduction*. McGraw-Hill Publishing Company.

Bordwell, D., & Thompson, K. (1985). "Fundamental Aesthetics of Sound in the Cinema". In Weis, E. & Belton, J. (eds), *Film Sound: Theory and Practice*. New York: Columbia University Press.

Branigan, E. (1989). "Sound and Epistemology in Film". *The Journal of Aesthetics and Art Criticism* 47(4), 311–324.

Buhler, J., Neumeyer, D., and Deemer, R. (2010). *Hearing the Movies: Music and Sound in Film History*. New York: Oxford University Press.

Bull, M. (2018). *The Routledge Companion to Sound Studies*. Abingdon, Oxon: New York, NY: Routledge.

Burch, N. (1985). "On the Structural Use of Sound". In E. Weis and J. Belton (eds), *Film Sound: Theory and Practice* (pp. 200–209). New York: Columbia University Press.

Burch, N. (1982). "Narrative/Diegesis—Thresholds, Limits". *Screen* 23(2), 16–33.

Carrigy, M. (2009a). "Bhuvan Shome". In Gopalan, L. (ed.), *The Cinema of India* (pp. 138–147). 24 Frames Series. London: Wallflower Press.

Carrigy, M. (2009b). "Meghe Dhaka Tara". In Gopalan, L. (ed.), *The Cinema of India* (pp. 126–137). 24 Frames Series. London: Wallflower Press.

Chatterjee, S. A. (2020). "Meghe Dhaka Tara". *Upperstall Magazine*. https://upperstall.com/film/meghe-dhaka-tara/ (accessed 4 July 2022).

Chakrabarty, D. (2007), *Provincializing Europe: Postcolonial Thought and Historical Difference*. Princeton: Princeton University Press.

Chatterjee, P. (1993), *The Nation and its Fragments: Colonial and Postcolonial Histories*. Princeton: Princeton University Press.

Chatterjee, S. (2012a, June 7). "Movie Review Shanghai". *NDTV Movies.com*. http://movies.ndtv.com/movie-reviews/movie-review-shanghai-715 (accessed 4 July 2022).

Chatterjee, S. (2012b, June 8). "Movie Review Shanghai". *The Sunday Indian*. www.thesundayindian.com/en/story/movie-review-shanghai/112/36105/ (accessed 4 July 2022).

Chattopadhyay, B. (2007). "Sound Memories: In Search of Lost Sounds in Indian Cinema". *Journal of the Moving Image* 6, 102–111. Jadavpur University Press.

Chattopadhyay, B. (2012a). "Sonification of Cinema: Studying the Use of Location Sound in Indian Films" (conference paper). *Music & The Moving Image Conference*, NYU.

Chattopadhyay, B. (2012b). "Sonic Menageries: Composing the Sound of Place". *Organised Sound* 17(3), 223–229.

Chattopadhyay, B. (2012c). "Certain Observations on the Evolution of Sound in Indian Cinema". *CPH PIX Catalogue*. http://cphpix.dk/article/368 (accessed 13 April 2023).

Chattopadhyay, B. (2013a). "Auditory Situations: Notes from Nowhere". *Journal of Sonic Studies* 4. https://www.researchcatalogue.net/view/269178/269179 (accessed 10 September 2023).

Chattopadhyay, B. (2013b). *Elegy for Bangalore*. Frankfurt am Main: Gruenrekorder.

Chattopadhyay, B. (2013c). "Interview". In Lane, C. & Carlyle, A. (eds.), *In the Field: The Art of Field Recording*, pp. 49 – 59. London: Uniformbooks.

Chattopadhyay, B. (2013d). "The Cinematic Soundscape: Conceptualising the Use of Sound in Indian Films". *SoundEffects* 2(2), 66–78.

Chattopadhyay, B. (2014a). "Object-Disoriented Sound: Listening in the Post-Digital Condition". *A Peer-Reviewed Journal About* 3(1). http://www.aprja.net/?p=1839 (accessed 13 April 2023).

Chattopadhyay, B. (2014b). "The Invisible Sound: A Study in the Trajectories of Sound Practice in Indian Films by Online Archival Research". *IASA Journal* 42, 89–95.

Chattopadhyay, B. (2014c). "Sonic Drifting: Sound, City and Psychogeography". *SoundEffects* 3(3), 138–152.

Chattopadhyay, B. (2015). "The Auditory Spectacle: Designing Sound for the 'Dubbing Era' of Indian Cinema". *The New Soundtrack* 5(1), 55–68. Edinburgh: Edinburgh University Press.

Chattopadhyay, B. (2016). "Being There: Evocation of the Site in Contemporary Indian Cinema". *Journal of Sonic Studies* 12. https://www.researchcatalogue.net/view/286592/286593 (accessed 10 September 2023).

Chattopadhyay, B. (2017). "Reconstructing Atmospheres: Ambient Sound in Film and Media Production". *Communication and the Public* 2(4), 352–364. London: SAGE Publication.

Chattopadhyay, B. (2018). "Orphan Sounds: Locating Historical Recordings in Contemporary Media". *Organised Sound* 23(2), 181–188. Cambridge: Cambridge University Press.

Chattopadhyay, B. (2020). "Post-immersion: Towards a Discursive Situation in Sound Art". *RUUKKU Studies in Artistic Research* 13. https://www.researchcatalogue.net/view/555994/555995 (accessed 10 September 2023).

Chattopadhyay, B. (2021a). *The Auditory Setting: Environmental Sounds in Film and Media Arts*. Edinburgh: Edinburgh University Press.

Chattopadhyay, B. (2021b). *Between the Headphones: Listening to the Practitioner* (2021b). Newcastle upon Tyne: Cambridge Scholars Publishing.

Chattopadhyay, B. (2022a). *Sound Practices in the Global South: Co-Listening to Resounding Plurilogues*. London: Palgrave Macmillan.

Chattopadhyay, B. (2022b). "Room Ambience: Home as Heard in Film and Media Arts". *Journal of Sonic Studies* 22. https://www.researchcatalogue.net/view/1440447/1440448 (accessed 10 September 2023).

Chattopadhyay, B. (2023a). "An Equal Sound". *Rewire Festival Context Programme* 2023. www.rewirefestival.nl/feature/an-equal-sound (accessed 5 April 2023).

Chattopadhyay, B. (2023b). "Re-Sounding Souths". *CTM Festival Journal*. www.ctm-festival.de/magazine/re-sounding-souths (accessed 5 April 2023).

Chattopadhyay, B. (2023c). "Dhvāni: Resonance". In Herzogenrath, B. (ed.), *Concepts: A Travelogue*, pp 113–120. New York: Bloomsbury.

Chattopadhyay, B. (2024). *Sonic Perspectives from the Global South: Connecting Resonances*. New York: Bloomsbury (forthcoming).

Chaudhuri, A. (2013, August). "Satyajit Ray: A Retrospective". *Financial Times*. www.ft.com/content/d4b1d302-fb82-11e2-8650-00144feabdc0 (accessed 4 July 2022).

Chaudhuri, S. (2007). "Space, Interiority and Affect in *Charulata* and *Ghare Baire*". *Journal of the Moving Image* 6. http://jmionline. org/article/space_interiority_and_affect_in_charulata_and_ghare_baire (accessed 3 July 2022).

Chhabra, A. (2014, February 14). "Review: *Highway* is Extremely Disappointing". *Rediff.com*. www.rediff.com/movies/report/review-highway-is-extremely-disappointing/20140214.htm (ccessed 3 July 2022).

Chion, M. (1991). "Quiet Revolution…And Rigid Stagnation". Translated by Ben Brewster. *Rendering the Real* 58 (Autumn): 69–80.

Chion, M. (1994). *Audio-Vision: Sound on Screen*. Translated and edited by C. Gorbman. New York: Columbia University Press.

Chion, M. (2009). *Film, a Sound Art*. New York: Columbia University Press.

Chuter, J. (2013, April 19). "Review: Budhaditya Chattopadhyay – *Elegy for Bangalore*". *ATTN:Magazine*. www.attnmagazine.co.uk/review-budhaditya-chattopadhyay-elegy-for-bangalore/ (accessed 3 July 2022).

Clair, R. (1929/1985). "The Art of Sound". In Weis, E. & Belton, J. (eds), *Film Sound: Theory and Practice*, pp 92–95. New York: Columbia University Press.

Clark, G. (1982). "Review of *Disco Dancer*". www.thespinningimage.co.uk/cultfilms/displaycultfilm.asp?reviewid=11568 (accessed 4 July 2022).

Cooper, D. (2000). *The Cinema of Satyajit Ray: Between Tradition and Modernity*. New York: Cambridge University Press.

Doane, M. A. (1980). "The Voice in the Cinema: The Articulation of Body and Space". In Altman, R. (ed.), *Yale French Studies, No. 60, Cinema/Sound*, pp. 33–50. New Haven, CT: Yale University Press.

Doane, M. A. (1985). "Ideology and the Practice of Sound Editing and Mixing". In E. Weis and J. Belton (eds), *Film sound: Theory and Practice*, pp. 54–62. New York: Columbia University Press.

Deleuze, G. and Guattari, F. (1972/2004). *Anti-Œdipus*. Translated by Robert Hurley, Mark Seem and Helen R. Lane. London and New York: Continuum.

D'Escriván, J. (2009). "Sound Art (?) on/in Film". *Organised Sound* 14(1), 65–73.

Demers, J. (2010). *Listening through the Noise: The Aesthetics of Experimental Electronic Music*. New York: Oxford University Press.

Dissanayake, W. and Sahai, M. (1992). *Sholay: A Cultural Reading*. New Delhi: Wiley Eastern Ltd.

Dore, B. (2017, June 7). "Payal Kapadia: Over the Clouds". *Open the Magazine*. https://openthemagazine.com/cinema/payal-kapadia-over-the-clouds/ (accessed 1 March 2023).

Drever, L. J. (2002). "Soundscape Composition: The Convergence of Ethnography and Acousmatic Music". *Organised Sound* 7(1), 21–27.

Dumouchel, P. (2015). 'Mirrors of Nature: Artificial Agents in Real Life and Virtual Worlds'. In Cowdell, S., Fleming, C. and Hodge, J. (eds), *Mimesis, Movies, and Media: Violence, Desire, and the Sacred*. New York: Bloomsbury Academic.

Ďurovičová, N. (2003). "Local Ghosts: Dubbing Bodies in Early Sound Cinema". *Moveast* 9. http://www.filmintezet.hu/uj/kiadvanyok/moveast/moveast_9/durovicova.htm (accessed 1 March 2023).

Dyson, F. (2014). *The Tone of Our Times: Sound, Sense, Economy, and Ecology*. Cambridge, Massachusetts: The MIT Press.

Dyson, F. (2009). *Sounding New Media: Immersion and Embodiment in the Arts and Culture*. California: University of California Press.

Eisenberg, A. J. (2015). "Space". In Novak, D. & Sakakeeny, M. (eds), *Keywords in Sound*, pp. 193–207. Durham: Duke University Press.

Eisenstein, S, Pudovkin, V. I. and Alexandrov, G. V. (1985) "A Statement". In Weis, E. and Belton, J. (eds), *Film Sound: Theory and Practice*, pp. 83–85. New Haven: Yale University Press.

Findlay-Walsh, I. (2018). "Sonic Autoethnographies: Personal Listening as Compositional Context". *Organised Sound* 23(1), 121–130. Cambridge: Cambridge University Press.

Fischer, L. (1985). "Applause: The Visual and Acoustic Landscape". In Weis, E. & Belton, J. (eds), *Film Sound: Theory and Practice*, pp. 233–250. New York: Columbia University Press.

Foucault, M. (1980). "The Confession of the Flesh" interview. In Gordon, C. (ed.), *Power/Knowledge: Selected Interviews and Other Writings*, pp. 194–228. New York: Pantheon Books.

Ganguly, S. (2001). *Satyajit Ray: In Search of the Modern*. New Delhi: Indialog.

Garga, B.D. (2007). *From Raj to Swaraj: The Non-Fiction Film in India*. Delhi; Penguin India.

Gess, N. (2015). "Ideologies of Sound: Longing for Presence from the Eighteenth Century until Today." *Journal of Sonic Studies* 10. https://www.researchcatalogue.net/view/220291/220292 (accessed 7 September 2023).

Ghosh, H. (2012). "Notes on the use of sound in *Shanghai*."
FutureWork. https://m.facebook.com/note.php?note_id=382123985181336&_ft_ (accessed 4 July 2022).

Gokulsing, K. M. and Dissanayake, W. (2004). *Indian Popular Cinema: A Narrative of Cultural Change*. London: Trentham Books.

Gooptu, S. (2011). *Bengali Cinema: An Other Nation*. London and New York: Routledge.

Gopalakrishnan, A. (2009). "Before Resul Pookutty, there was Adoor Gopalakrishnan Rajaneesh Vilakudy". *Rediff.com*. www.rediff.com/movies/2009/mar/02adoor-gopalakrishnan-on-resul-pookutty-sound.htm (accessed 4 July 2022).

Gopalan, L. (2002). *Cinema of Interruptions: Action Genres in Contemporary Indian Cinema*. London: British Film Institute.

Gorbman, C. (1987). *Unheard Melodies: Narrative Film Music*. Bloomington: Indiana University Press.

Grimshaw, M. (2011). *Game Sound Technology and Player Interaction: Concepts and Developments*. Hershey: Information Science Reference, IGI Global.

Grimshaw, M. (2014). *The Oxford Handbook of Virtuality*. Oxford: Oxford University Press.

Gumbrecht, H. U. (2004). *Production of Presence: What Meaning Cannot Convey*. Stanford: Stanford University Press.

Ganti, T. (2012). *Producing Bollywood: Inside the Contemporary Hindi Film Industry*. Durham, North Carolina, United States: Duke University Press.

Hayward, S. (2006). *Cinema Studies: The Key Concepts*. London: Routledge.

Holman, T. (1997/2002/2010). *Sound for Film and Television*. Boston: Focal Press.

Ihde, D. (2007). *Listening and Voice: Phenomenologies of Sound*. New York: Sunny Press.

IMDb (Internet Movie Database) (2009). "Review". http://www.imdb.com/title/tt1093370/reviews (accessed 1 May 2023).

Iniarra, M. (2010, July). "Interview with Vishal Gondal, CEO of Indiagames Ltd". *TED Global, Oxford UK*.

Jakimowicz, M. (2010, October 13). "Art Review". *Deccan Herald*.

Jordan, R. (2012). "The Ecology of Listening while Looking in the Cinema: Reflective Audioviewing in Gus Van Sant's *Elephant*". *Organised Sound* 17(3), 248–256.

Kaul, M. (1977). "Communication". Symposium on THE CINEMA SITUATION, with Kumar Shahani, Adoor Gopalakrishnan, Shyam Benegal and Dileep Padgaonkar. https://theseventhart.files.wordpress.com/2011/07/the-cinema-situation_74.pdf (accessed 4 July 2022).

Kaul, M. (1983). "Towards a Cinematic Object". *Indian Cinema Super Bazaar*. https://theseventhart.files.wordpress.com/2011/07/towards-a-cinematic-object_mk.pdf (accessed 4 July 2022).

Kaul, M. (1991). "Seen from Nowhere". In Vatsavan, Kapita (ed.), *Concepts of Space: Ancient and Modern*. New Delhi: Abhinav Publications.

Kassabian, A. (2013). "The End of Diegesis as We Know It?" In John Richardson, Carol Vernallis, and Claudia Gorbman (eds), *The Oxford Handbook of New Audiovisual Aesthetics*, pp. 89-106. Oxford: Oxford University Press.

Kaleebullah, E. (2016, March 22). "Making Bollywood Movies With Dolby Atmos". *Gadgets 360*. www.gadgets360.com/tv/features/making-bollywood-movies-with-dolby-atmos-570913 (accessed 1 May 2023)

Khan, A., Puri, O., Benegal, D. Choudhury, M., Balsavar, A. et al. (2002). "Sync Sound and Indian Cinema". *Upperstall Magazine*. http://budhaditya.org/wp-content/uploads/2016/07/Sync-Sound-and-Indian-Cinema-Upperstall.Com_1.pdf (accessed 5 July 2023).

Kania, A. (2009). "Realism". In P. Livingston and C. Plantinga (eds), *The Routledge Companion to Philosophy and Film*, pp. 237–247. London and New York: Routledge.

Kelman, Ari Y. (2010). 'Rethinking the soundscape: A critical genealogy of a key term in sound studies'. *Senses and Society* 5/2: 212–234.

Kerins, M. (2011). *Beyond Dolby (Stereo): Cinema in the Digital Sound Age*. Bloomington: Indiana University Press.

Kerins, M. (2006). "Narration in the Cinema of Digital Sound". *The Velvet Light Trap* 58 (Fall), 41–54.

Kim-Cohen, S. (2009). *In the Blink of an Ear*. New York: Bloomsbury.

Kittler, F. (1999). *Gramophone, Film, Typewriter*. Translated by Geoffrey Winthrop-Young and Michael Wutz. Stanford, CA: Stanford University Press.

Kuhn, A. and Westwell, G. (2014). *A Dictionary of Film Studies*, in *Oxford Reference Online*. www.oxfordreference.com/view/10.1093/acref/9780199587261.001.0001/acref-9780199587261-e-0659# (accessed 5 July 2023).

LaBelle, B. & Martinho, C. (eds). (2011). *Site of Sound: Of Architecture and the Ear: Vol 2*. Berlin: Errant Bodies Press.

LaBelle, B. (2010). *Acoustic Territories: Sound Culture and Everyday Life*. New York: Continuum.

LaBelle, B. (2006/2015). *Background Noise: Perspectives on Sound Art*. New York: Bloomsbury Academic.

Lastra, J. (2000). *Sound Technology and the American Cinema: Perception, Representation, Modernity*. New York: Columbia University Press.

Lawrence, A. (1992). "Women's Voices in Third World Cinema". In Altman, R. (ed.), *Sound Theory/Sound Practice*, pp. 178–190. New York: Routledge.

Lee, K. M. (2004). "Presence, Explicated". *Communication Theory* 14(1), 27–50.

Lefebvre, H. (1992). *The Production of Space*. Hoboken, New Jersey: Wiley

LoBrutto, V. (1994). *Sound-on-Film: Interviews with Creators of Film Sound*. Westport: Praeger.

Lombard, M. and Ditton, T. (2006). *At the Heart of It All: The Concept of Presence, Journal of Computer-Mediated Communication* 3(2). https://academic.oup.com/jcmc/article/3/2/JCMC321/4080403 (accessed 10 September 2023).

Maitra, P. (ed.) (1995). *100 Years of Cinema*. Calcutta: Nandan.

McAdams, S. & Bigand, E. (1993). *Thinking in Sound: The Cognitive Psychology of Human Audition*. New York: Oxford University Press.

Mazumdar, R. (2009). "Satya". In Gopalan, L. (ed.), *The Cinema of India*, pp. 236–245. 24 Frames Series. London: Wallflower Press.

Messerli, D. (2011, December). "Trapped in the Past". *International Cinema Review*. http://internationalcinemareview.blogspot.dk/2011/12/trapped-in-past-by-douglas-messerli.html (accessed 5 July 2023).

Miller, J. and Glassner B. (2011). "The "Inside" and the "Outside": Finding Realities in Interviews". In David Silverman (ed.), *Qualitative Research: Issues of Theory, Method and Practice*, pp. 125–139. London: Sage.

Morcom, A. (2016). *Hindi Film Songs and the Cinema*. Farnham, UK: Ashgate Publishing.

Morley, D. and Robins, K. (1995). *Spaces of Identity: Global Media, Electronic Landscapes and Cultural Boundaries*. London: Routledge.

Mukherjee, M. (2012). *Aural Films, Oral Cultures: Essays on Cinema from the Early Sound Era*. Jadavpur: Jadavpur University Press.

Mukherjee, M. (2007). "Early Indian Talkies: Voice, Performance and Aura". *Journal of the Moving Image* 6, 39–61. Kolkata: Jadavpur University Press.

Mukim, M. (2017). "The Private Eye: Mani Kaul and the Cinema to Come". *Caravan Magazine*. https://caravanmagazine.in/reviews-essays/mani-kaul-cinema-come (accessed 7 September 2023).

Nafiu, R. F. (2018). "Goutam Ghose on the Future of Cinema". *Dhaka Tribune*. https://archive.dhakatribune.com/showtime/2018/04/20/goutam-ghose-future-cinema (accessed 1 March 2023).

Nancy, J. L. (2007). *Listening*. Translated by Charlotte Mandell. New York: Fordham University Press.

Novak, D., and Sakakeeny, M. (eds). (2015). *Keywords in Sound*. Durham; London: Duke University Press.

O'Donnell, E. (2004). "'Woman' and 'Homeland' in Ritwik Ghatak's films: Constructing Post-Independence Bengali Cultural Identity". *Jump Cut: A Review of Contemporary Media* 47. www.ejumpcut.org/archive/jc47.2005/ghatak/text.html (accessed 1 March 2023).

Papadomanolaki, M. (2013, October 3). "Review of *Elegy for Bangalore*, Gruenrekorder (Budhaditya Chattopadhyay, 2013)". *Field Reporter/Sonic Field*. https://sonicfield.org/310/ (accessed 1 March 2023).

Pinch, T. and Bijsterveld, K. (2013). *The Oxford Handbook of Sound Studies*. Oxford: Oxford University Press.

Pemmaraju, G. (2013). "Soundbaazi: The Sound of More than Music". *ArtConnect: The IFA Magazine* 7(1), 62–82.

Percheron, D. (1980). "Sound in Cinema and its Relationship to Image and Diegesis". In Altman, R. (ed.), *Yale French Studies Number 60: Cinema/Sound*, pp. 16–23. New Haven: Yale University Press.

Plantinga, C. (2009). "Emotion and Affect". In Carl R. Plantinga and Paisley Livingston (eds.), *The Routledge Companion to Philosophy and Film*, pp. 356–365. London: Routledge.

Prince, S. (1996). "True Lies: Perceptual Realism, Digital Images, and Film Theory". *Film Quarterly* 49(3), 27–37.

Rajadhyaksha, A. (2007). "An Aesthetic for Film Sound in India?" *Journal of the Moving Image* 6. https://www.jmionline.org/article/an_aesthetic_for_film_sound_in_india (accessed 7 September 2023).

Rajadhyaksha, A. (2009). *Indian Cinema in the Time of Celluloid: From Bollywood to the Emergency*. New Delhi: Indiana University Press.

Rajadhyaksha, A. (2016). *Indian Cinema: A Very Short Introduction*. Oxford: Oxford University Press

Ray, B. (2011). *Amader Katha*. Calcutta: Ananda Publishers.

Ray, S. (1976). *Our Films, Their Films*. Delhi: Orient Blackswan.

Ray, S. (2011). *Satyajit Ray on Cinema*. Edited by Sandip Ray. New York: Columbia University Press.

Ray, S. (1978, April). "Seminar Speech at American Film Institute", April 1978. https://youtu.be/TdFi1noweog (accessed 13 May 2023).

Reiter, U. (2011). "Perceived Quality in Game Audio". In Grimshaw, M. (ed.), *Game Sound Technology and Player Interaction: Concepts and Developments*. Hershey: Information Science Reference.

Reuters (2009, February 23). "Resul Pookutty Wins Sound Mixing Oscar for 'Slumdog'". www.reuters.com/article/idINIndia-38161420090223 (accessed 13 May 2023).

Robinson, A. (1989). *Satyajit Ray: The Inner Eye*. California: University of California Press.

Robinson, D. (2020). *Hungry Listening: Resonant Theory for Indigenous Sound Studies*. Durham, North Carolina: Duke University Press.

Rogers, A. (2013). *Cinematic Appeals: The Experience of New Movie Technologies*. New York: Columbia University Press.

Sengupta, A. (2007). "Seeing through the Sound: Certain Tendencies of the Soundtrack in Satyajit Ray's Films of the 1970s". *Journal of the Moving Image* 6, 86–101.

Sen, R. (2012, June 8). "Review: *Shanghai*, a Frighteningly Fine Film". *Rediff.com*. www.rediff.com/movies/review/review-shanghai-a-frighteningly-finefilm/20120608.htm (accessed 13 May 2023).

Sergi, G. (2004). *The Dolby Era: Film Sound in Contemporary Hollywood*. Manchester: Manchester University Press.

Seton, M. (2003). *Portrait of a Director: Satyajit Ray*. New Delhi: Penguin Books India.

Skalski, P. & Whitbred, R. (2010). "Image versus Sound: A Comparison of Formal Feature Effects on Presence and Video Game Enjoyment". *PsychNology Journal* 8(1), 67–84.

Sonnenschein, D. (2001). *Sound Design: The Expressive Power of Music, Voice, and Sound Effects in Cinema*. Michigan: Michael Wiese Productions.

Schafer, R. M. (1994). *The Soundscape: Our Sonic Environment and the Tuning of the World*. Rochester: Destiny Books.

Shaviro, S. (2010). *Post Cinematic Affect*. Ropley, UK: John Hunt Publishing.

Sobchack, V. (2005). "When the Ear Dreams: Dolby Digital and the Imagination of Sound". *Film Quarterly* 58(4), 2–15.

Sobchack, V. (1991). *The Address of the Eye: A Phenomenology of Film Experience*. Princeton, NJ: Princeton University Press.

Sonnenschein, D. (2001). *Sound Design: The Expressive Power of Music, Voice, and Sound Effects in Cinema*. Studio City, CA: Michael Wiese Productions.

Sorfa, D. (2014). "A Glossary of Film Terms". www.academia.edu/1739157/A_ Glossary_of_Film_Terms (accessed 13 May 2023).

Steingo, G. & Sykes, J. (2019). *Remapping Sound Studies*. Durham: Duke University Press.

Sterne, J. (2003). *The Audible Past: Cultural Origins of Sound Reproduction*. Durham: Duke University Press.

Sterne, J. (2006). "The Death and Life of Digital Audio". *Interdisciplinary Science Reviews* 31(4), 338–348.

Sterne, J. (2012). "Sonic Imaginations". In Sterne, J. (ed.), *The Sound Studies Reader*, pp. 1–17. London: Routledge.

Sterne, J. (ed.) (2012). *The Sound Studies Reader*. London: Routledge.

Stubblefield, T. (2006). "Ritwik Ghatak and the Role of Sound in Representing Post-Partition Bengal". *Post Script: Commerce* 25(3), 17-29.

Tan, E. S. (2011). *Emotion and the Structure of Narrative Film: Film as an Emotion Machine*. New York and London: Routledge.

Théberge, P., Devine, K., and Everrett, T. (2015). *Living Stereo: Histories and Cultures of Multichannel Sound*. New York: Bloomsbury.

Times, H. (2010, July 24). "35 Years On, the Sholay Fire Still Burns". www.hindustantimes.com/bollywood/35-years-on-the-sholay-fire-still-burns/story-mcY7jLPKbKvYCwfrC9h4JP.html (accessed 4 July 2022).

Thompson, E. (2004). *The Soundscape of Modernity: Architectural Acoustics and the Culture of Listening in America, 1900–1933*. London: The MIT Press.

Tilak, S. G. (2017, December 27). "A German Cinematographer's Love Affair with Indian Cinema". BBC News Services. www.bbc.com/news/world-asia-india-42422599 (accessed 1 March 2023).

Times of India (2013, August 22). "Tamil Leads as India Tops Film Production", *Times of India*. https://timesofindia.indiatimes.com/city/hyderabad/tamil-leads-as-india-tops-film-production/articleshow/21967065.cms (accessed 1 May 2023).

Tybjerg, C. (2013). "The Sown and the Waste – Or, The Psychedelic Writing of Film History". *NECSUS_European Journal of Media Studies*, 2(2), 509–525.

Waller, D. A. and Nadel, L., (eds). (2013). *Handbook of Spatial Cognition*, pp. 81–97. Washington DC: American Psychological Association.

Weiss, A. S. (2011). *Varieties of Audio Mimesis: Musical Evocations of Landscape*. Berlin: Errant Bodies Press.

Vasudevan, R. (2001). "Nationhood, Authenticity and Realism in Indian Cinema: The Double Take of Modernism in the Work of Satyajit Ray". *Journal of the Moving Image* 2, 52–76.

Vijayakar, R. (2009). *The History of Indian Film Music: A Showcase of the Very Best in Hindi Cinema*. New Delhi: Times Group Books.

Voegelin, S. (2010). *Listening to Noise and Silence: Towards a Philosophy of Sound Art*. London: Bloomsbury Academic.

Filmography

1942: A Love Story (Vidhu Vinod Chopra 1994)
36 Chowringhee Lane (Aparna Sen, 1981)

Aapothkalin Trikalika ('The Kali of Emergency', Ashish Avikunthak, 2016)
A Night of Knowing Nothing (Payal Kapadia, 2021)
ABCD: Anybody Can Dance (Remo D'Souza, 2013)
Afternoon Clouds (Payal Kapadia, 2017)
Ajantrik (Ritwik Ghatak, 1958)
Albert Pinto Ko Gussa Kyon Aata Hai ('Why Does Albert Pinto Get Angry?', Saeed Akhtar Mirza, 1980)
All That Breathes (Shaunak Sen, 2022)
Amar Akbar Anthony (Manmohan Desai, 1977)
Ankur (Shyam Benegal, 1974)
Antarjali Jatra (Goutam Ghose, 1987)
Apu Trilogy (Satyajit Ray, 1955–1959)
Aranyer Din Ratri ('Days and Nights in the Forest', Satyajit Ray, 1970)
Around the World (Raj Kapoor, 1967)
Article 15 (Anubhav Sinha, 2019)
Arvind Desai Ki Ajeeb Dastaan ('The Strange Fate of Arvind Desai', Saeed Akhtar Mirza, 1978)
Asha Jaoar Majhe ('Labour of Love', Aditya Vikram Sengupta, 2014)
Ashad Ka Ek Din ('One Day in the Rainy Season', Mani Kaul, 1971)
Awara (Raj Kapoor, 1951)

Bagh Bahadur (Buddhadeb Dasgupta, 1989)
Bareilly Ki Barfi ('*Bareilly's Barfi*', Ashwiny Iyer Tiwari, 2017)
Bariwali (Rituparno Ghosh, 1999)
Bhaag Milkha Bhaag (Rakeysh Omprakash Mehra, 2013)
Bhagya Chakra ('Wheel of Fate', Nitin Bose, 1935)
Bhagya Chakra ('Wheel of Fate', Bose, 1935a)
Bhumika (Shyam Benegal, 1977)
Bhuvan Shome (Mrinal Sen, 1969)
Bobby (Raj Kapoor, 1973)
Bombay (Mani Ratnam, 1995)
Blindfold (Binoy Karamen, 2023)

Calcutta '71 (Mrinal Sen, 1972)
Calcutta Trilogy (Mrinal Sen, 1971, 1972, and 1973)

Chaowa Pawa ('To Want and to Receive', Yatrik, 1959)
Charachar (Buddhadeb Dasgupta, 1993)
Chokher Bali (Rituparno Ghosh, 2003)
Chowringhee (Pinaki Mukherjee, 1968)
Chhuti (Arundhati Devi, 1967)
Court (Chaitanya Tamhane, 2014)

Daud (Ram Gopal Varma, 1997)
Dear Zindagi (Gauri Shinde, 2016)
Dedh Ishqiya ('One and a Half Passionate,' Abhishek Chaubey, 2014)
Deepar Prem (Arundhati Devi, 1983)
Dekha (Goutam Ghose, 2001)
Delhi-6 (Rakeysh Omprakash Mehra, 2009)
Deya Neya (Sunil Bandyopadhyay, 1963)
Dharti Ke Lal (K.A. Abbas, 1946)
Dhobi Ghat (Kiran Rao, 2011)
Dhoom 3 (Vijay Krishna Acharya, 2013)
Dhoop Chhaon ('Sun and Shade', Nitin Bose, 1935)
Dhoop Chhaon ('Sun and Shade', Bose, 1935b)
Dhrupad (Mani Kaul, 1983)
Dil Se... ('From the Heart', Mani Ratnam, 1998)
Disco Dancer (B. Subhash, 1982)
Do Bigha Zamin (Bimal Roy, 1953)
Dooratwa (Buddhadeb Dasgupta, 1978)
Dum Maaro Dum (Ramesh Sippy, 2011)
Dweepa ('The Island', Girish Kasaravalli, 2002)
Dwelling (Kabir Mohanty, 2006)
Duvidha (Mani Kaul, 1973)

Gandhi (Richard Attenborough, 1982)
Gangs of Wasseypur (Anurag Kashyap, 2012)
Ghanchakkar (Raj Kumar Gupta, 2013)
Ghawre Bairey Aaj (Aparna Sen, 2019)
Ghatashraddha ('The Ritual', Girish Kasaravalli, 1979)
Gokul (Arundhati Devi, 1985)
Gunga Jumna (Nitin Bose, 1961)

Haider (Vishal Bhardwaj, 2014)
Handheld (Kabir Mohanty, 2009)

Hey Ram (Kamal Haasan, 2000)
Highway (Imtiaz Ali, 2014)

In Memory (Kabir Mohanty, 2009, 2012)
Interview (Mrinal Sen, 1970)

Kaagaz Ke Phool ('Paper Flowers', Guru Dutt, 1959)
Kaalpurush (Buddhadeb Dasgupta, 2008)
Kahani ('The Story', Sujoy Ghosh, 2012)
Kalyug (Shyam Benegal, 1981)
Kaminey ('Scoundrels', Vishal Bhardwaj, 2009)
Karz (Subhash Ghai, 1980)
Khandhar ('The Ruins', Mrinal Sen, 1984)
Krrish 3 (Rakesh Roshan, 2013)
Krrish (Rakesh Roshan, 2006)
Killa ('The Fort', Avinash Arun, 2015)
Komal Gandhar ('E-Flat', Ritwik Ghatak, 1961)

Lagaan (Ashutosh Gowariker, 2001)
Lal Darja (Buddhadeb Dasgupta, 1997)
Love Sex aur Dhokha (Dibakar Banerjee, 2010)

Madras Cafe (Shoojit Sircar, 2013)
Mahanagar ('The Big City', Satyajit Ray, 1963)
Mahaprithibi ('World Within, World Without', Mrinal Sen, 1991)
Maine Pyar Kiya ('I Have Loved', Sooraj Barjatya, 1989)
Mandi (Shyam Benegal, 1983)
Manthan (Shyam Benegal, 1976)
Mard ('Macho', Manmohan Desai, 1985)
Margarita with a Straw (Shonali Bose, 2015)
Masaan ('Fly Away Solo', Neeraj Ghaywan, 2015)
Mathilukal ('The Walls', Adoor Gopalakrishnan, 1990)
Maya Darpan (Kumar Shahani, 1972)
Megh o Roudra (Arundhati Devi, 1969)
Meghe Dhaka Tara ('The Cloud-Capped Star', Ritwik Ghatak, 1960)
Mondo Meyer Upakhyan (Buddhadeb Dasgupta, 2002)
Monsoon Wedding (Mira Nair, 2001)
Mother India (Mehboob Khan, 1957)
Mr India (Kapur, 1987)

Mr. & Mrs. '55 (Guru Dutt, 1955)
Mr. and Mrs. Iyer (Aparna Sen, 2002)
Mughal-E-Azam (K. Asif, 1960)

Nagarik (Ritwik Ghatak, 1952)
Nainsukh (Amit Dutta, 2010)
Nautanki Saala! (Rohan Sippy, 2013)
Naya Daur (B.R. Chopra, 1957)
Nayak ('Hero', Satyajit Ray, 1966)
Nazar ('The Gaze', Mani Kaul, 1991)
Neecha Nagar (Chetan Anand, 1946)
Nishant (Shyam Benegal, 1975)

October (Shoojit Sircar, 2018)
Om-Dar-Ba-Dar (Kamal Swaroop, 1988)
Oridathu (G. Aravindan, 1987)

Padatik ('The Foot Soldier', Mrinal Sen, 1973)
Padi Pishir Barmi Baksha (Arundhati Devi, 1972)
Padma Nadir Majhi (Goutam Ghose, 1992)
Paroma (Aparna Sen, 1984)
Paromitar Ek Din (Aparna Sen, 2000)
Pather Panchali ('Song of the Little Road', Satyajit Ray, 1955)
Pratham Basanta (Dinen Gupta, 1971)
Pyaasa (Guru Dutt, 1957)

Ra.One (Anubhav Sinha, 2011)
Rangeela (Ram Gopal Varma, 1995)
Roja ('Rose', Mani Ratnam, 1992)
Rudaali ('Female Weeper', Kalpana Lajmi,1993)

Saagar ('The Sea', Sippy, 1985)
Salaam Bombay! (Mira Nair, 1988)
Samapti in *Teen Kanya* (*'Three Daughters'*, Satyajit Ray, 1961)
Sati (Aparna Sen, 1989)
Scene of Crime (Amar Kanwar, 2011)
Shahid (Hansal Mehta, 2013)
Shahenshah (Anand, 1988)
Shair (M.S. Chawla, 1949)
Shanghai (Dibakar Banerjee, 2012)

Ship of Theseus (Anand Gandhi, 2015)
Sivaji 3D (Shankar, 2012)
Sholay ('Embers', Ramesh Sippy, 1975)
Shubho Mahurat (Rituparno Ghosh, 2003)
Shree 420 (Raj Kapoor, 1955)
Siddheshwari (Mani Kaul, 1989)
Sivaji: The Boss (S. Shankar, 2007)
Sonata (Aparna Sen, 2017)
Sonchidi (Amit Dutta, 2011)
Subarnarekha ('The Golden Thread', Ritwik Ghatak, 1965)
Such a Morning (Amar Kanwar, 2017)
Suno Baranari ('Listen O Lady', Ajoy Kar, 1960)
Swapner Din (Buddhadeb Dasgupta, 2005)

Tahader Katha (Buddhadeb Dasgupta, 1993)
Tarang (Kumar Shahani, 1984)
Taxi Driver (Chetan Anand, 1954)
Thalaivaa ('Leader', A.L. Vijay, 2013)
The Last Lear (Rituparno Ghosh, 2007)
The Namesake (Mira Nair, 2006)
The Reluctant Fundamentalist (Mira Nair, 2012)
The Sky Is Pink (Shonali Bose, 2019)

Unishe April (Rituparno Ghosh, 1994)
Uski Roti ('Daily Bread', Mani Kaul, 1969)
Utsab (Rituparno Ghosh, 2000)
Uttara (Buddhadeb Dasgupta, 2000)

Vishwaroopam (Kamal Hasan, 2013)

Waqt ki Awaz ('The Sound of Time', Bapayya, 1988)

Yugant (Aparna Sen, 1995)

Index

For Product Safety Concerns and Information please contact our EU representative GPSR@taylorandfrancis.com
Taylor & Francis Verlag GmbH, Kaufingerstraße 24, 80331 München, Germany

www.ingramcontent.com/pod-product-compliance
Ingram Content Group UK Ltd.
Pitfield, Milton Keynes, MK11 3LW, UK
UKHW011002180926
475701UK00004B/514
* 9 7 8 1 0 4 1 1 8 6 3 9 7 *